Surface Activity

Principles, Phenomena, and Applications

Series in *Polymers, Interfaces, and Biomaterials*

Other books in the series:

Alexander Grosberg, Editor, *Theoretical and Mathematical Models in Polymer Research: Modern Methods in Polymer Research and Technology* (1998).

Teruo Okano, Editor, *Biorelated Polymers and Gels: Controlled Release and Applications in Biomedical Engineering* (1998).

Also available:

Alexander Grosberg and Alexei R. Khokhlov, *Giant Molecules: Here, There, and Everywhere* (1997).

Jacob Israelachvili, *Intermolecular and Surface Forces, Second Edition* (1992).

Surface Activity
Principles, Phenomena, and Applications

Kaoru Tsujii
TOKYO RESEARCH CENTER
KAO CORPORATION
TOKYO, JAPAN

ACADEMIC PRESS
San Diego London Boston
New York Sydney Tokyo Toronto

This book is printed on acid-free paper. ♾

ACADEMIC PRESS
525 B Street, Suite 1900, San Diego, CA 92101, USA
1300 Boylston Street, Chestnut Hill, MA 02167, USA
http://www.apnet.com

United Kingdom Edition published by
ACADEMIC PRESS LIMITED
24-28 Oval Road, London NW1 7DX
http://www.hbuk/co.uk/ap/

Library of Congress Cataloging-in-Publication Data

Tsujii, Kaoru.
Surface activity : principles, phenomena, and applications / Kaoru
Tsujii.
p. cm. — (Polymers, interfaces, and biomaterials)
Includes bibliographical references and index.
ISBN 0-12-702280-5 (alk. paper)
1. Surface active agents. 2. Surface chemistry. I. Title.
II. Series.
TP994.T88 1998
668′.1—dc21 97-51306
CIP

Printed in the United States of America
98 99 00 01 02 IC 9 8 7 6 5 4 3 2 1

Contents

Preface by Series Editor

Surfaces and interfaces are a concept of fundamental importance in materials science and technology. All materials have surfaces. When two or more materials are put together, interfaces appear. In composite materials, the interfaces play a crucial role in determining their mechanical, optical, and electronic properties.

The surfaces and interfaces can be drastically altered when surfactant molecules are applied to the area. They not only modify the interactions between two materials at the interface, but also the bulk properties of the composite. Surfactants are a key material in processes involving adhesion, coating, mixing, domain formations, and many other phenomena in medical, pharmaceutical, chemical, and electronics industries.

Surfactants exhibit a variety of interesting phenomena such as formations of micelles, vesicles, layers, and gels. It is a fascinating example of self-assembling. Surfactants have many phases and undergo phase transitions between them. Understanding these phenomena is a great challenge for physics and chemistry and is of increasing technological importance.

Dr. Kaoru Tsujii is one of the pioneers and leaders in science and technology of surfactants. He has long explored the field, both theoretically and in industrial applications. This book illustrates the fundamentals and principles behind these phenomena and their applications to our everyday life and industry. It will serve as an excellent text and reference for researchers and students who wish to know more about surfactants, one of the key materials in twenty-first century technology.

Toyoichi Tanaka, Series Editor
Massachusetts Institute of Technology
March, 1998

Preface by Author

Surface activity is an important phenomenon in our daily life as well as in many kinds of industry. So much phenomena are governed by surface activity that understanding its principles, phenomena, and applications can help one to live wisely and to work creatively. Some examples of familiar products that are surface-active materials are the soap and shampoo used on your body and the detergent and fabric softener used on your clothing. When you cook, you may use a surface-modified (Teflon-coated) frying pan that can be easily cleaned up; this too is a product of the technology of surface activity.

The scientific field of surface activity is now in a new stage of progress. Technology that will enable us to arrange or assemble molecules as we want will be one of the most far-reaching technologies of the twenty-first century. For instance, the concept of molecular electronics perfectly depends on this technology. Already we have the technology to assemble atoms to form molecules in terms of (synthetic) organic chemistry due to the contributions of quantum mechnanics, which provides the principles for this chemistry. The next epoch will be technology for artificially assembling molecules to make something useful in our social life. Thus, the science of surface activity—the only technique at present for molecular assembling or construction—is now in the spotlight.

This book is written to contribute to the up-to-date understandings of the science and technology of surface activity from basic principles to practical applications. The book is intended for company research scientists and engineers, university professors and graduate students, and even undergraduate students who are working or want to work in the field of surface and colloid science. Distinctive features of this book are (1) it is written with one logical philosophy from basic principles to applications, (2) basic principles, interesting phenomena, and useful applications of surface activity are discussed clearly and can be easily understood even by beginners in this field, and (3) potential future applications—such as lyotropic liquid crystals, liposomes and vesicles, bilayer membranes, and LB-films—are discussed together with conventional ones.

The author would like to express his sincere gratitude to Professor Toyoichi

Tanaka of MIT for providing the chance to write this book. He also appreciates Dr. Zvi Ruder, Ms. Elizabeth Voit, and Abby Heim of Academic Press for their kind help in publishing the book. He is indebted to his wife, Yukiko, for her continuous encouragement during the writing of this book.

Kaoru Tsujii
Tokyo, Japan
March, 1998

Surface Activity

Principles, Phenomena, and Applications

Chapter 1 Surface Activity and Surface and Interfacial Tension

The concept of surface activity is very closely related with surface and interfacial tension of a material, and thus cannot be understood without first understanding what surface and interfacial tension are. So, let us start the first chapter with a description of the origin of surface and interfacial tension.

1.1 Origin of Surface and Interfacial Tension

1.1.1 SURFACE TENSION

Condensed matters (liquids and solids) have surface tension because cohesive energy is present between their molecules. A molecule in the bulk of a condensed matter interacts with attractive force with its surrounding molecules. For example, a water molecule in bulk liquid phase makes some (at most four) hydrogen bonds as well as van der Waals interactions, and a carbon atom in a diamond crystal has four C–C covalent bonds with nearest-neighbor atoms. Molecules present at a surface, however, cannot fully form such bonding and/or interaction since they have no (or few) interacting molecule in the vacuum (vapor) side and thus have excess energy compared with those in the bulk phase. This excess energy existing in the surface molecules or atoms is defined as *surface tension.* Figure 1.1 shows a schematic representation (a) of surface molecules that have some unconnected hydrogen bonds and less van der Waals interaction at the surface of liquid water and (b) of dangling bonds of carbon atoms in a diamond crystal.

One can readily imagine along this line that any molecules present on an edge or a corner would have more excess energy than those at the surfaces. These excess energies may be called *line tension* for an edge, and *point tension* for a corner. Line tension is actually recognized as a factor that determines the shape of small liquid droplets or the contact angle (Princen, 1969; Scheludko *et al.,* 1976; Pethica, 1977; de Feijter, 1988) as well as the meniscus shape of thin liquid films

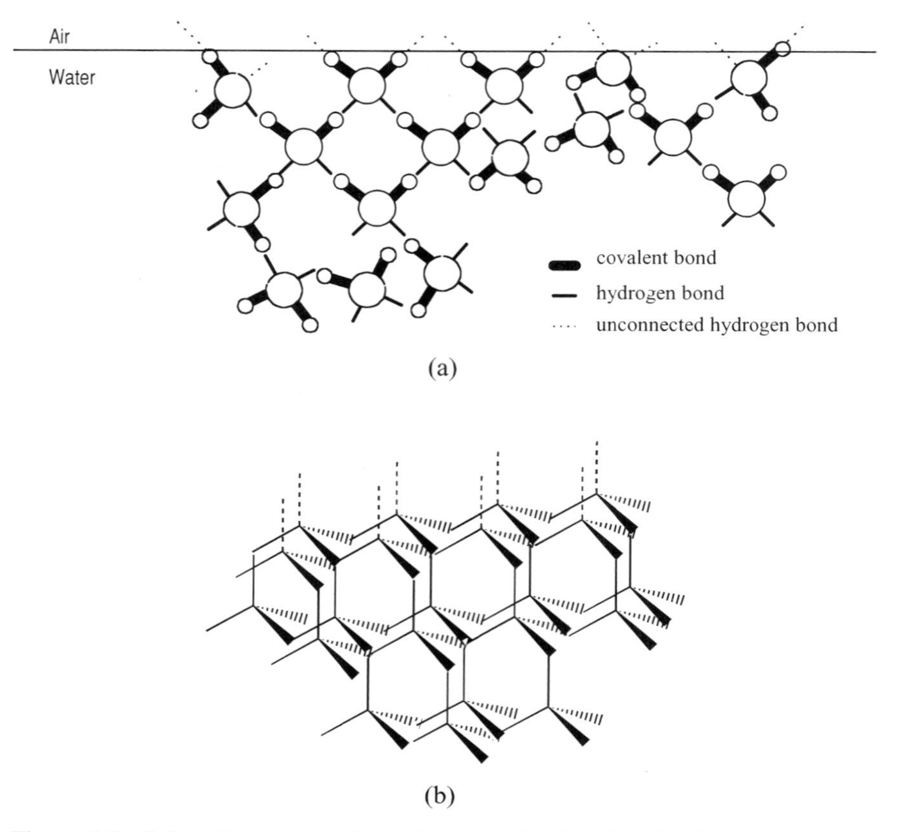

Figure 1.1 Schematic representations of water molecules (a) and carbon atoms of diamond crystals (b) present at the surfaces. Surface molecules or atoms have some unconnected bonds (denoted as dotted lines in the figures) and/or cannot have full interactions with surrounding molecules or atoms; thus they have excess free energy (surface tension).

(Ivanov and Kralchevsky, 1988). Dislocation defects in metal crystals are also a typical example of line tension. In addition, mechanical properties and plastic deformations of metals and alloys can be successfully explained in terms of the line tension of dislocations (Friedel, 1964; Granato *et al.,* 1964). A point tension, however, has not yet been observed experimentally in any kind of phenomena; it is just a conceptual quantity at present.

Surface tension minimizes the surface area of liquid matter to reduce the surface excess energy. A typical experiment to show this is soap film spread on a frame. Figure 1.2 exhibits this experiment. One side of the rectangular frame is movable, and a soap film is spread inside the frame. If one stretches the mobile frame bar and then releases it, the soap film shrinks spontaneously to minimize its surface area. Designating the force that pulls the frame as f and the length of the frame bar as l, we can define the force per unit length as

$$\gamma = \frac{f}{2l}. \tag{1.1}$$

The quantity γ is the surface tension. We multiply l in the denominator by 2 since two surfaces (top and bottom) are present in the soap film. When the movable bar is pulled against the surface tension to the length of x, the work (w) done is fx, and the area (s) increased by this expansion is $2lx$. Then the surface tension (γ) can be rewritten as

$$\gamma = \frac{fx}{2lx} = \frac{w}{s}. \tag{1.2}$$

Eq. (1.2) means that the surface tension is excess free energy per unit area. Thus, the surface tension can be expressed as both force/length and free energy/area.

As we easily see from the definition of the surface tension, materials that have higher cohesive energy must have higher surface tension. Solid matter has high cohesive energy and thus high surface tension. On the other hand, gas has no surface tension because molecules in gas phase have no cohesive energy. Table 1.1 shows some data of surface tension of metals in solid and liquid states. Of course, we see that the surface tension of a metal in the solid state is higher than that of the same metal in a liquid state because of the higher cohesive energy.

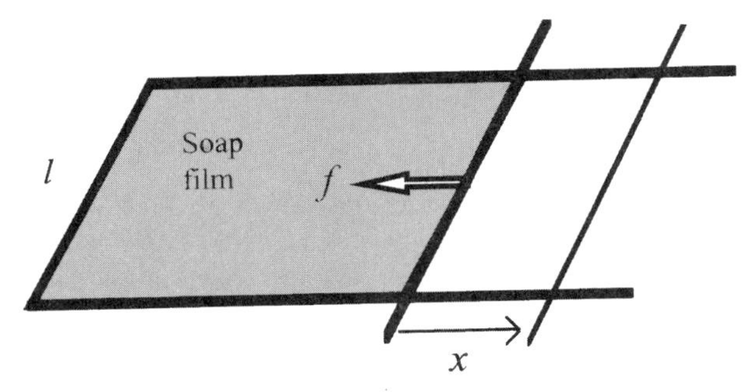

Figure 1.2 Simple experiment to show that surface tension makes the surface area minimum. A soap film spread inside the rectangular frame pulls the mobile frame bar by the force of f.

Table 1.1 **Surface Tension of Metals in Solid and Liquid (in Parentheses) States**

Metal	*Temperature (°C)*	*Solid or (Liquid)*	*Surface Tension (mNm^{-1})*
Gold	1300 (1120)	Solid (Liquid)	1400 or 1510 (1128)
Silver	900 (995)	Solid (Liquid)	1140 (923)
Copper	1050 (1140)	Solid (Liquid)	1430 or 1670 (1120)
Iron	1400 (1530)	Solid (Liquid)	1670 (1700*)
Tin	150 (700)	Solid (Liquid)	704 (538)
Aluminum	(700)	(Liquid)	(900)
Mercury	(20)	(Liquid)	(476)

Data for the liquid state were taken from Bondi (1953). Data for the solid state were taken from Udin (1952).

*Datum was obtained by extrapolating the data of steels to zero alloy and carbon concentration, and might contain larger experimental error.

Surface tension (and interfacial tension) data of popular liquids frequently used in a laboratory are listed in Table 1.2. Water shows a high value since it has high cohesive energy due to hydrogen bond formation.

Surface tension changes with temperature: it decreases with increasing temperature. This is because the cohesive energy between molecules becomes smaller and smaller with increasing temperature, causing surface tension to decrease

Table 1.2 **Surface Tension of Popular Solvents and Their Interfacial Tension with Water** (Davies and Rideal, 1963)

Solvent	*Temperature (°C)*	*Surface Tension (mNm^{-1})*	*Interfacial Tension with Water (mNm^{-1})*
Water	20	72.8	—
Water	25	72.0	—
Bromobenzene	25	35.75	38.1
Benzene	20	28.88	35.0
Benzene	25	28.22	34.71
Toluene	20	28.43	
n-Octanol	20	27.53	8.5
Chloroform	20	27.14	
Carbon tetrachloride	20	26.9	45.1
n-Octane	20	21.8	50.8
Diethylether	20	17.01	10.7

and finally disappear at a critical point. Figure 1.3 shows the surface tension versus temperature curve of chloroform (Adamson, 1982a).

1.1.2 INTERFACIAL TENSION

Some tension or excess free energy is present also at the interfaces between two condensed matters. Figure 1.4 shows a schematic representation of intermolecular cohesive forces at the surface (upper figure) and the interface (lower figure) between two liquids A and B. When two surfaces of liquid A and B are separated each other, they have, of course, their own surface tensions γ_A and γ_B. When the two surfaces come into contact each other to form an interface, an attractive force between molecules of A and B appears. This attractive force partially compensates the excess free energy of the molecules present at the surfaces of A and B, and consequently reduces the surface tension of two liquids.

Denoting the attractive cohesive energy between A and B molecules as σ_{AB}, we can write an equation for the interfacial tension between both liquids:

$$\gamma_{AB} = \gamma_A + \gamma_B - 2\sigma_{AB} \tag{1.3}$$

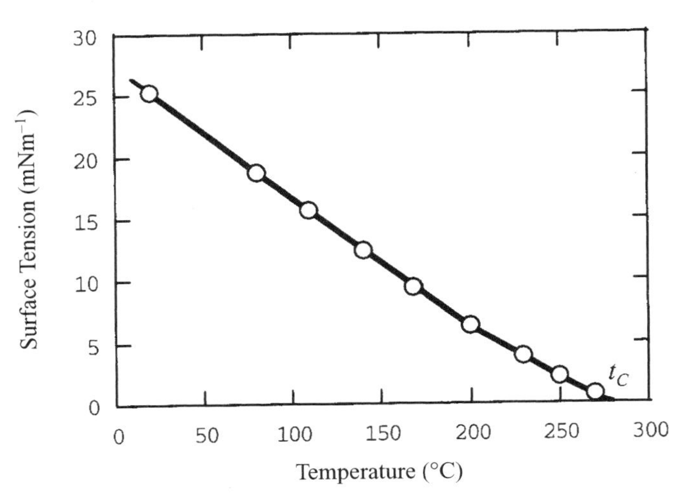

Figure 1.3 Surface tension of chloroform against temperature modified from "Physical Chemistry of Surfaces," 4th ed., A. W. Adamson, copyright ©1982. Reprinted by permission of John Wiley & Sons, Inc. Surface tension decreases with increasing temperature and disappears at the critical point (t_C).

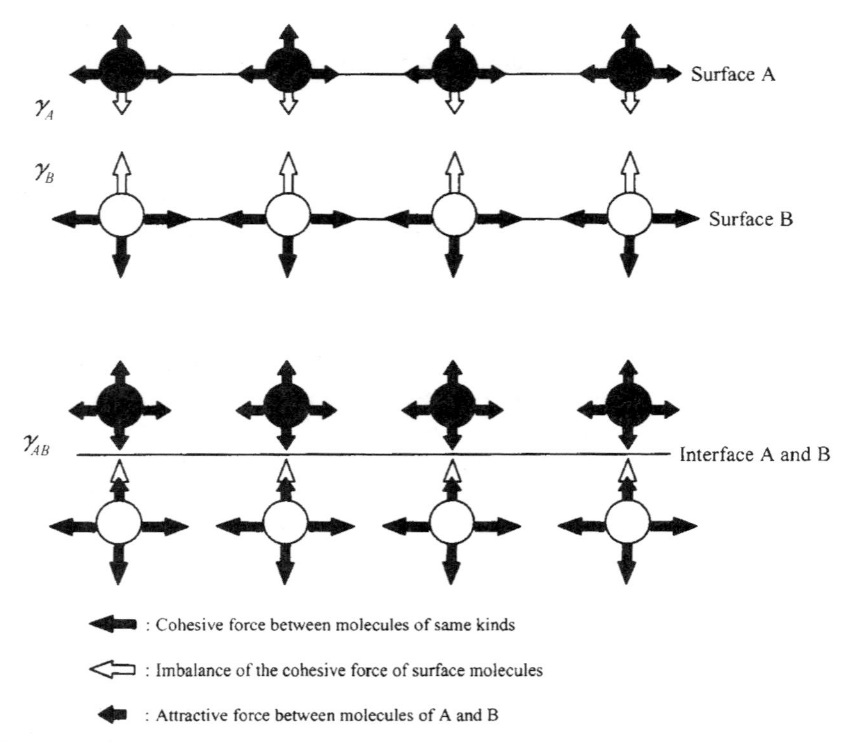

Figure 1.4 Schematic representation of intermolecular cohesive forces at the surface (upper figure) and the interface (lower figure) between two liquids A and B. The attractive force between A and B causes the cohesive energy σ_{AB}, which partially compensates the surface tensions of liquids A and B.

If the liquids A and B possess like interaction—i.e., if the interactions between the A molecules themselves are the same kind as those between the B molecules themselves (say, van der Waals attraction), then σ_{AB} can be expressed as $(\gamma_A \gamma_B)^{1/2}$ (Israelachvili, 1991). This relationship allows us to calculate the interfacial tension from the surface tension values of individual two liquid phases. As a special case, if σ_{AB} is equal to γ_A the interfacial tension between both liquids is just the difference of the two surface tensions of the liquids, i.e., $\gamma_{AB} = \gamma_B - \gamma_A$. This is often the case when the two liquids are water (liquid B) and a nonpolar organic solvent (liquid A). The attractive interaction between water and such an organic solvent is similar to that between the organic molecules themselves: Both are van der Waals attractions. The interfacial tension, γ_{AB}, originates in this case from the imbalance of the cohesive energy due to hydrogen bonds between water molecules, which cannot be compensated by the interaction with the organic molecules. Referring to Table 1.2, we can see this in the data of interfacial tensions of water with carbon tetrachloride and *n*-octane.

It is worth noting that the interfacial tensions between water and some kinds of alcohol are generally low. Alcohol molecules orient at the interface with the hydrophilic (–OH) group toward the water phase and make the interaction (mainly hydrogen bonding) with water molecules. This interaction makes the σ_{AB} value larger and the γ_{AB} lower. If the σ_{AB} value is much larger than both values of γ_A and γ_B, then the γ_{AB} becomes negative. Negative interfacial tension makes the area of interface larger. Gradually the two kinds of liquid penetrate each other in as small a size as possible and may finally reach the molecular mixture of both components. This result is, of course, a true solution. Thus, strong interaction between A and B makes the interface disappear and forms a solution.

1.2 What Is Surface Activity?

Surface tension of liquid (say, water) varies with the dissolution of some compounds into the liquid. Figure 1.5 shows the surface tension of aqueous solutions of some compounds as a function of their concentrations (Padday, 1969). We can see that inorganic salts increase the surface tension of water, but methanol makes it lower. When the surface property is remarkably changed by the addition of a small amount of some compound, the phenomenon is called *surface activity*. Surface activity in the most typical case appears in surface tension lowering, and the compounds that depress the surface tension are called *surface active compounds* or *active substances*. A surface active agent (abbreviated as *surfactant*) is the most powerful substance that lowers the surface tension of the solvent (water in particular).

The change of surface tension by dissolution of some compounds originates from the adsorption of the solute onto the surface of the solutions. This relationship is given by the *Gibbs' adsorption isotherm*, expressed as

$$-\frac{d\gamma}{d\mu} = -\frac{1}{RT}\frac{d\gamma}{d\ln a} = \Gamma, \tag{1.4}$$

where γ is the surface tension of the solution; μ and a are the chemical potential and the activity of the solute in the solution, respectively; Γ is the adsorption amount at the surface; and R and T the gas constant and the absolute tempera-

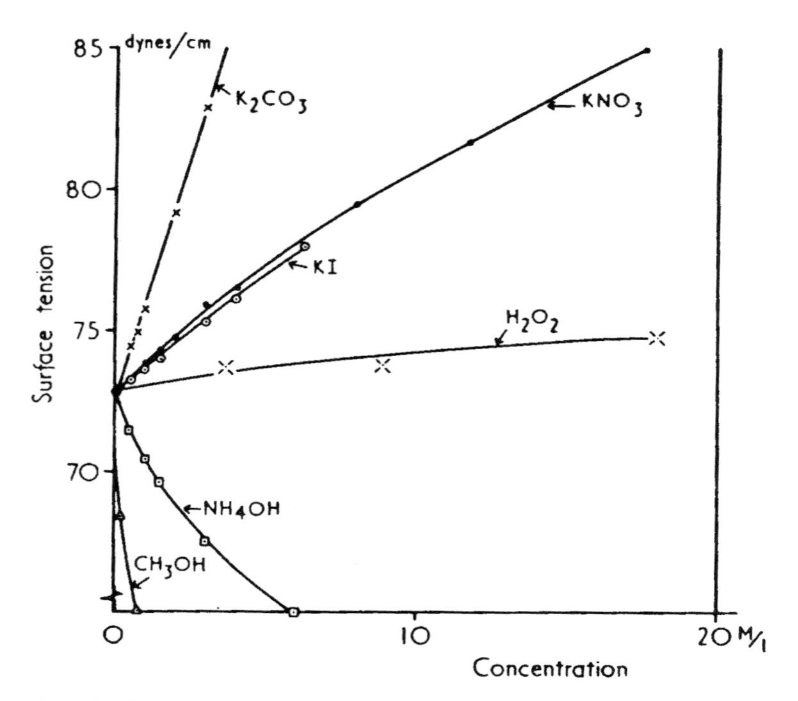

Figure 1.5 Surface tension of aqueous solutions of some compounds as a function of their concentrations. Reprinted from "Surface and Colloid Science" vol. 1, p. 92 by permission of Plenum Pub. Corp.

ture, respectively. When C, the concentration of the solute, is low enough, Eq. (1.4) can be rewritten as

$$-\frac{1}{RT}\frac{d\gamma}{d\ln C}=\Gamma. \tag{1.5}$$

Eq. (1.5) indicates that more surface active solute depresses the surface tension more effectively (by less concentration) and thus adsorbs in greater amounts onto the surface of the solution. So the Gibbs' adsorption isotherm can be regarded as a quantitative representation of the surface activity. The above indication is quite reasonable from the thermodynamic point of view since the adsorption of solute actually occurs owing to the free energy gain (decrease) by this adsorption process. The surface tension lowering is, of course, the free energy gain of this case. Eq. (1.5) gives us an interesting result. When surface tension is lowered with the addition of a compound, the adsorption amount Γ of the compound is positive; but when the surface tension is elevated with the addition of a compound, Γ must be negative. The meaning of negative adsorption is that the concentration of the compound at the surface of the solution is lower than that in the bulk liquid; this will be made clear later.

In deriving Eq. (1.5), we tacitly assume that the solvent does not adsorb at the surface of the solution. This assumption implies that the concentrations of solvent and solute are as shown in Figure 1.6. Both concentrations of solute and solvent change, of course, abruptly at the surface of the solution.

We select a geometric plane as the interface between the solution and the vapor phases at which total mass of the solvent is balanced, assuming that the solvent concentrations of both phases are kept constant until this plane (see Figure 1.6). In general, the concentration change of the solute at the surface of the solution behaves differently from that of the solvent. So if we calculate the total amount of solute under the same assumption for the solvent (i.e., that the concentration in bulk phase is maintained until this plane), we have an excess amount of the solute in this case (shaded area in Figure 1.6). This surface excess is the adsorption amount Γ that appeared in Eqs. (1.4) and (1.5).

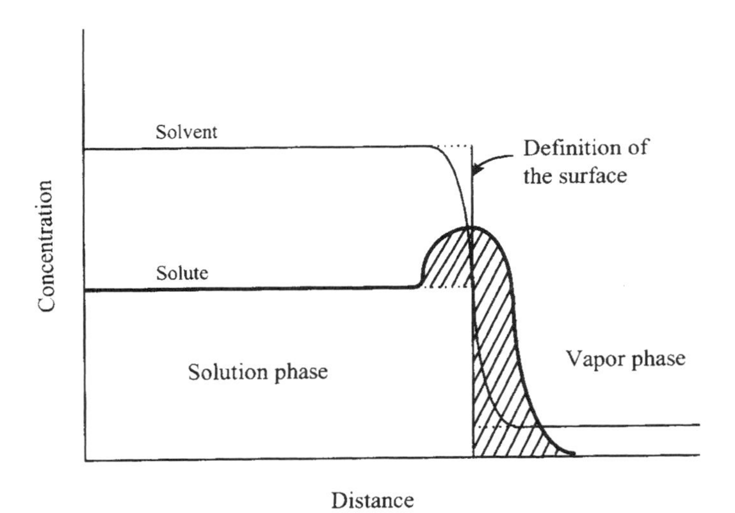

Figure 1.6 Concentration profiles of both solvent and solute at the surface of the solution. The surface is commonly defined as a geometric plane at which no solvent adsorbs. When the surface is so defined, solute usually shows some excess amounts (positive or negative). This surface excess is called the *adsorption amount* and is shaded here. (It is positive adsorption in this case.)

When we have a multicomponent system, the Gibbs' adsorption isotherm is written in the more general form

$$-d\gamma = RT\, \Sigma_i\, \Gamma_i\, d \ln C_i, \tag{1.6}$$

where Γ_i and C_i are the surface excess and the concentration of the *i*th component, respectively. One can select any component as a solvent and assume its surface excess to be zero. The Gibbs' adsorption equation in multicomponent systems of electrolytic solutes is somewhat more complicated; details are given in the literature (Tajima, 1971; Ikeda, 1977).

1.3 Surface Activity in Aqueous Systems and Hydrophobic Interaction

1.3.1 BEHAVIOR OF HYDROCARBONS IN AQUEOUS SOLUTIONS

Water molecules can make hydrogen bonds with each other and gain large cohesive energy. Thus, a water molecule prefers to have other water molecules as nearest neighbors so they will be stabilized by the hydrogen bonding. Hydrocarbons, on the other hand, are inert in polar interactions and can have only van der Waals interactions. As a consequence, hydrocarbons are not welcome by water molecules and tend to escape from them. In other words, hydrocarbons are insoluble in water.

However, the hydrocarbons can be dissolved into water with the aid of attached hydrophilic polar groups. What happens? The hydrocarbon groups escape contact with the water molecules by adsorbing passively at the surface of the solution. If some hydrophobic solid or liquid (oil) is present in the solution, the hydrocarbon groups also adsorb onto the surface of the hydrophobic materials. Consequently, hydrocarbons are very surface active in aqueous systems.

Another way hydrocarbons escape contact with water molecules is to gather together among themselves; this is the *aggregation,* or *association, phenomenon.* Formations of higher-order structure of proteins, biological membranes, and micelles of surfactant are typical examples of this phenomenon.

One can now understand that the adsorption and/or the aggregation of hydrocarbons take place passively due to the strong cohesive energy between the water molecules. This apparent attractive interaction between hydrocarbon groups that are segregated from water is called *hydrophobic interaction.* Hydrophobic interaction works in aqueous systems because the dissolved molecule is inert in polar interactions including hydrogen bonding. Therefore, it must work in aqueous solutions of any compounds that have no polar group. Rare gases are another example of such a solute.

As pointed out previously (Section 1.2), adsorption of solute takes place due to the decrease of surface tension of the solutions by the adsorption process. We have also seen that hydrophobic interaction promotes the adsorption of nonpolar solutes. One may conclude from these two propositions that hydrophobic nonpolar solute must depress the surface tension of the solutions and this is actually the case. The high surface tension of water originates from the imbalance of large cohesive energy (chiefly the unconnected hydrogen bonds) of the molecules at the surface. When a nonpolar solute (say, hydrocarbon) adsorbs at the water surface, water molecule(s) present at the surface can migrate into the bulk water phase and make hydrogen bonds fully with surrounding water molecules. These newly formed hydrogen bonds result in the free energy gain (decrease) in the system and partially compensates the free energy imbalance of water molecules at the surface of the aqueous solution. This is the reason the hydrocarbon compounds can depress markedly the surface tension of water and can adsorb strongly.

1.3.2 HYDROPHOBIC INTERACTION AND ICEBERG FORMATION

When a nonpolar molecule is inserted into hydrogen bond networks of water phase, some of the hydrogen bonds of the water molecules surrounding the solute must be broken. This state must be at a higher free energy level and be favorable for the hydrophobic interaction. Since water molecules do not prefer such an unstable state of high free energy, they will rearrange themselves to form hydrogen bonding again as much as possible. Thus, hydrogen bond networks are formed around the nonpolar solutes (such as hydrocarbons). This process caused by hydrophobic solute is called *iceberg formation* (Frank and Evans, 1945).

It is clear from the above explanation that iceberg formation makes the hydrophobic interaction weaker. From Figure 1.7, which shows the solubility of some aromatic liquid hydrocarbons into water as a function of temperature (Shinoda, 1977; Shinoda and Becher, 1978a), one can see that the solubility starts to deviate to the upper side at the temperature of iceberg formation. At higher temperatures, molecular motion of water is vigorous and hydrogen bond formation is not favorable. Consequently, the solution behaves normally in the plot of $\ln X_2$ versus $1/T$ because of the random mixing solution of molecules in the system. At lower temperatures, however, water molecules can form hydrogen bonds around the solute molecules (iceberg formation), and the solubility of hy-

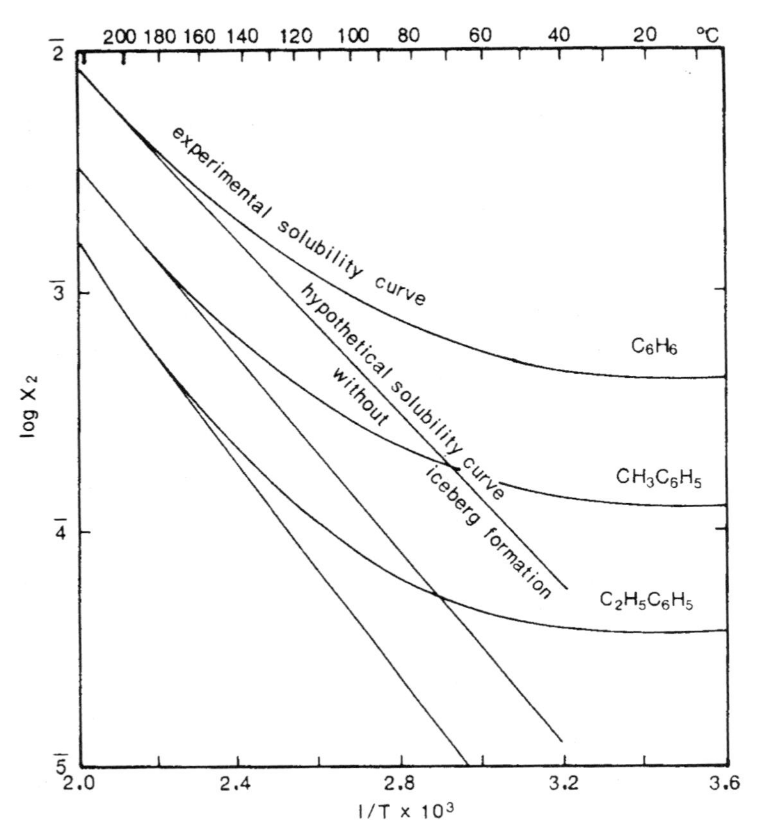

Figure 1.7 Solubility of some aromatic hydrocarbons into water as a function of temperature. Reprinted from "Principles of Solution and Solubility," p. 130 by courtesy of Marcel Dekker Inc.

drocarbons deviate to the higher side. The increase of hydrocarbon solubility results in the weak hydrophobic interaction.

As mentioned above, iceberg formation makes the hydrophobic interaction weaker. Some textbooks and papers, however, propose the completely opposite explanation—i.e., that iceberg formation is the origin of hydrophobic interaction (Kauzmann, 1959; Nemethy and Scheraga, 1962a, 1962b; Tanford, 1980a). Historically speaking, this opposing interpretation is given first and corrected to the above one (Shinoda, 1977; Shinoda and Becher, 1978a; Shinoda *et al.*, 1987; Privalov and Gill, 1989). Readers must take note of this point when discussing hydrophobic interaction and iceberg formation.

Refreshing Room!

GIBBS' EQUATION AND REVERSE OSMOSIS MEMBRANE

Gibbs' adsorption isotherm, Eq. (1.5), tells us that negative adsorption occurs in any compound that elevates the surface tension of water. Inorganic salts actually increase the surface tension of water, and their concentration at the surface of the solution must be lower than that in the bulk aqueous phase. Thus, we could expect that below the first or second molecular layer from the surface, the solution may be almost pure water even in the concentrated salt solutions. If we could slice off one or two molecular layers at the surface, would we obtain fresh water from the saline solution (say, sea water).

Realization of this fantastic but crucially sophisticated idea was attempted but did not succeed due to a technological problem. An inventor of the reverse osmosis membrane, Dr. Sourirajan, considered this method to obtain freshwater hopeless. He suggested instead that the negative adsorption of the inorganic salts at the interface between solid materials and aqueous saline solution be used (Sourirajan, 1970). If extremely tiny holes are made into which several molecular layers of water are able to adsorb, only pure water may pass through the holes. Loeb and Sourirajan selected cellulose acetate as the membrane material and obtained very hopeful results (Loeb and Sourirajan, 1958). Such reverse osmosis membranes have since been extensively applied in many fields of industry, such as desalination of seawater, concentration of liquid food, treatment of wastewater, and so on.

Gibbs' adsorption isotherm gives the same concept to all of us, but only the person who wishes to do something new can find something important for science and/or technology from an ordinary and popular theory!

Chapter 2 | Surface Active Substances

2.1 Characteristic Molecular Structure of Surface Active Compounds

A surface active molecule consists of two parts with opposing character—like Dr. Jekyll and Mr. Hyde. One part is hydrophilic and the other is hydrophobic. A typical surface active molecule, sodium octadecanoate (one of the main components of soap), is shown in Figure 2.1. As shown in the figure, surface active molecules are usually drawn like a match, with the head of a match being the hydrophilic group (which is the sodium carboxylate group in the case of soap) and the stem being the hydrophobic group. The hydrophobic group is usually a hydrocarbon chain, and so it is often called a lipophilic group. The substance that shows surface activity is called by several names—surface active substance, surface active compound, and surface active agent (surfactant). Since surface active substances exhibit affinity to both water and oils, they may also be called amphiphilic compounds, amphiphiles, or amphoteric compounds. We do not give an exact definition for the above terminology, but roughly distinguish them as follows. Surface active substances and surface active compounds are almost identical: They are substances affecting the surface or interfacial properties in the widest sense. The terms *amphiphilic* or *amphoteric* are used for compounds that have at least one hydrophilic and hydrophobic group in its molecule. We use the term *surfactant* in the narrowest sense: This is the compound that dramatically lowers the surface tension of water and forms aggregates like micelles in aqueous media.

There are many kinds of hydrophilic and hydrophobic groups of surfactants and/or amphiphilic compounds. Typical examples of the groups are listed in Figure 2.2. Hydrocarbon chains are the most popular hydrophobic group. Carbon numbers of the hydrocarbon chain are from 8–22, (usually from 10–18) in most of the surfactants. Every surfactant produced from natural fats or oils has a hydrocarbon chain with an even number of carbon atoms. In their chemical structures, some chains are saturated and some are unsaturated hydrocarbons; some are straight and some are branched. Some surfactants having two hydrocarbon

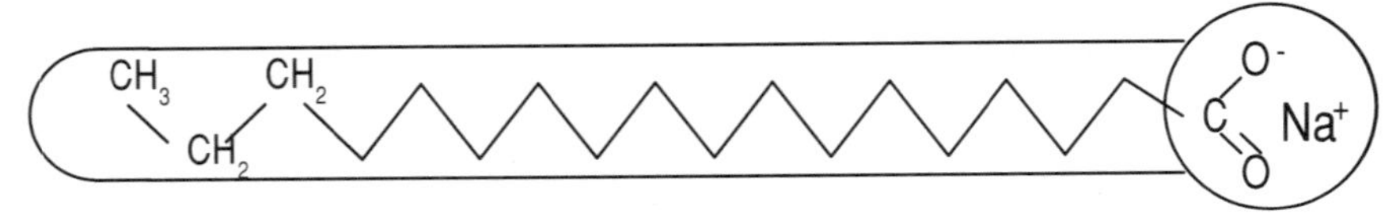

Figure 2.1 Molecular structure of sodium octadecanoate.

chains in one molecule easily form bilayer membranes and vesicles. Polycyclic hydrocarbon skeletons are also hydrophobic groups in some natural amphiphilic compounds such as bile salts and cholesterol. The biological and physicochemical significance of these very special structures in a hydrophobic group is not clear yet. The surfactants having fluorocarbon chains show the characteristic property of markedly lowering the surface tension of water due to weak cohesive energy between fluorocarbon groups. Polydimethylsiloxane (silicone) chains are not popular yet as a hydrophobic group, but may become much more important in the near future since silicone oils are now used extensively in cosmetic and/or toiletry products.

Surfactants are classified by their hydrophilic groups. Anionic, cationic, amphoteric, and nonionic surfactants contain anion, cation, amphoteric ion, and nonion, respectively, as their hydrophilic group. Anionic surfactants are mainly used as detergents. Cationic surfactants are particularly useful in fabric softeners and hair conditioners. Amphoteric surfactants are usually used as a booster to enhance the detergency and/or foaming of anionic surfactants. Nonionic surfactants, most popular as emulsifiers, also show strong detergency for oily dirt. Some surfactants (cholesterol, monoglycerides, monoglycerylether, saccharide derivatives, etc.) have only hydroxyl groups as the hydrophilic group. Most of these are water-insoluble, but sometimes show very interesting and characteristic properties.

2.2 Surfactants and Their Characteristic Properties

We mentioned the general characteristic features of surface active substances in the previous section. In this section, the properties and the characteristics of individual surface active agents (surfactants) will be described, together with their synthetic route. One can see easily from Figure 2.2 that there are many kinds of surfactant molecules that can be made by combinations of the hydrophobic and hydrophilic groups shown in the figure. Since it is impossible to describe every

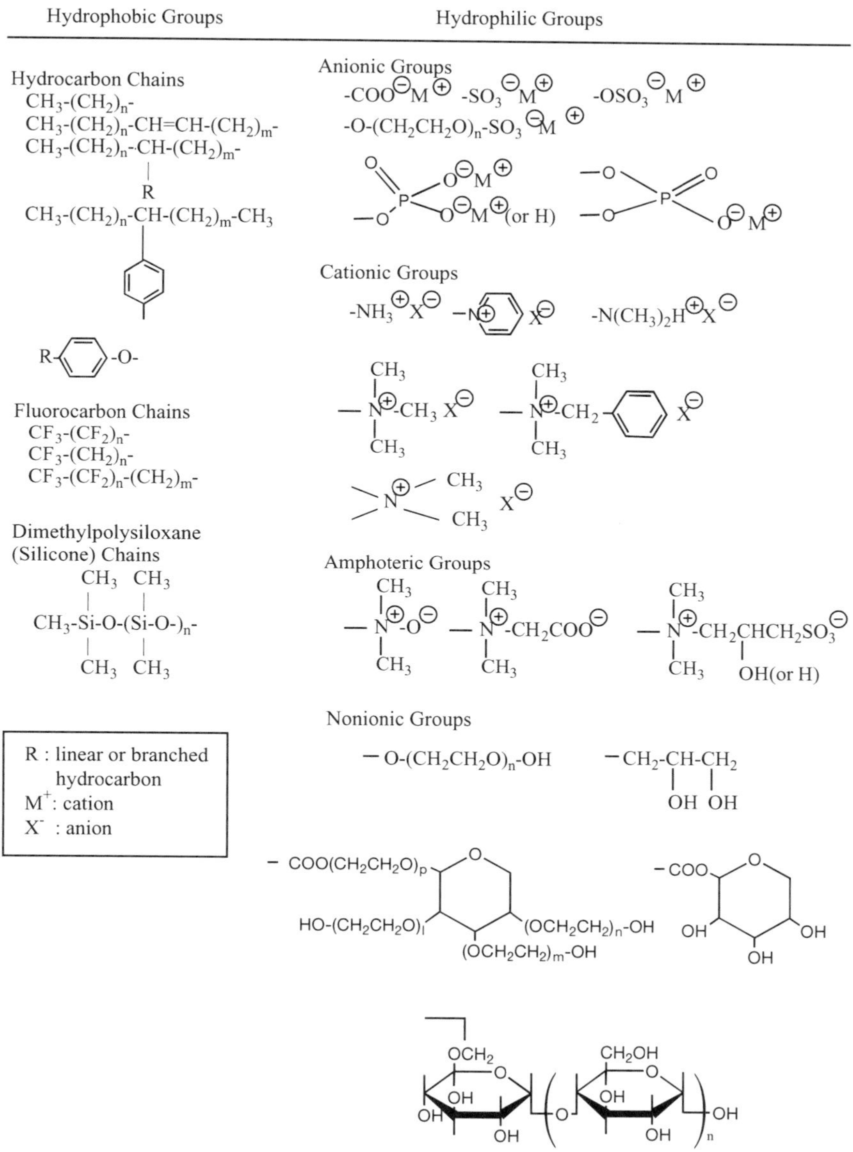

Hydrophobic Groups	Hydrophilic Groups
Hydrocarbon Chains $CH_3-(CH_2)_n-$ $CH_3-(CH_2)_n-CH=CH-(CH_2)_m-$ $CH_3-(CH_2)_n-CH(R)-(CH_2)_m-$ $CH_3-(CH_2)_n-CH(C_6H_4-)-(CH_2)_m-CH_3$ $R-C_6H_4-O-$	**Anionic Groups** $-COO^{\ominus}M^{\oplus}$ $-SO_3^{\ominus}M^{\oplus}$ $-OSO_3^{\ominus}M^{\oplus}$ $-O-(CH_2CH_2O)_n-SO_3^{\ominus}M^{\oplus}$ $-O-P(=O)(O^{\ominus}M^{\oplus})(O^{\ominus}M^{\oplus}$ (or H)) $(-O)_2P(=O)O^{\ominus}M^{\oplus}$ **Cationic Groups** $-NH_3^{\oplus}X^{\ominus}$ $-N^{\oplus}C_5H_5\ X^{\ominus}$ $-N(CH_3)_2H^{\oplus}X^{\ominus}$ $-N^{\oplus}(CH_3)_3\ X^{\ominus}$ $-N^{\oplus}(CH_3)_2-CH_2-C_6H_5\ X^{\ominus}$ $>N^{\oplus}(CH_3)_2\ X^{\ominus}$
Fluorocarbon Chains $CF_3-(CF_2)_n-$ $CF_3-(CH_2)_n-$ $CF_3-(CF_2)_n-(CH_2)_m-$	**Amphoteric Groups** $-N^{\oplus}(CH_3)_2-O^{\ominus}$ $-N^{\oplus}(CH_3)_2-CH_2COO^{\ominus}$ $-N^{\oplus}(CH_3)_2-CH_2CH(OH\text{(or H)})CH_2SO_3^{\ominus}$
Dimethylpolysiloxane (Silicone) Chains $CH_3-Si(CH_3)_2-O-(Si(CH_3)_2-O-)_n-$	**Nonionic Groups** $-O-(CH_2CH_2O)_n-OH$ $-CH_2-CH(OH)-CH_2(OH)$ $-COO(CH_2CH_2O)_p-$ (sorbitan ring with $HO-(CH_2CH_2O)_l$, $(OCH_2CH_2)_n-OH$, $(OCH_2CH_2)_m-OH$) $-COO-$ (sorbitan ring with OH, OH, OH) $-OCH_2$ (glucoside)$-(O-$glucose$)_n-OH$ (CH₂OH, OH, OH)

R : linear or branched hydrocarbon
M^+: cation
X^- : anion

Figure 2.2 Typical hydrophobic and hydrophilic groups of surfactants.

kind of surfactant, we will focus only on the most popular and typical ones. R and R′ in the chemical structure denote the hydrophobic groups consisting of saturated or unsaturated, straight or branched hydrocarbon chains. M^+ and X^- are cation and anion, respectively.

2.2.1 SYSTEMATIC FLOWCHART OF SURFACTANT SYNTHESIS

Although there are many kinds of surfactants, they are produced in systematic routes. The starting raw material is mainly fats or oils such as tallow, coconut, and palm oils. In some cases, they are petroleum hydrocarbons. Figure 2.3 shows a synthetic flowchart of major surfactants. The course of "fat → fatty acid → fatty alcohol → fatty amine" is the main route for surfactant synthesis. Many kinds of anionic, cationic, amphoteric, and nonionic surfactant are derived from these fatty acids, alcohols, and amines. As we describe the synthetic methods of the compounds in the main route, we will give the synthesizing reactions of each individual surfactant.

a. Production of Fatty Acids from Fats

There are currently two common methods for synthesizing fatty acids from fats. One is direct hydrolysis of fat with water vapor at high temperature (250–260°C) and high pressure (50–60 atm). The other is a methyl ester synthesis where the first step is an ester-exchange reaction, followed by hydrolysis of the methyl ester to fatty acids. The former is the more popular process, but the latter has an advantage of much easier fractional distillation of methyl esters than that of fatty acids themselves.

Enzymatic hydrolysis of fats by a lipase is not yet popular because the production cost is still high compared with the other two methods. However, this enzymatic production method is expected to grow as a future technology and is now used to produce some special fatty acids such as polyunsaturated fatty acids that are not stable at high temperatures and pressures.

b. Synthesis of Fatty Alcohols from Fatty Acid Methyl Esters

This reductive reaction is done under the conditions of high temperatures and high pressures of hydrogen gas (250–300°C, 250–300 atm), usually with copper-chromium catalysts. Direct reduction of fatty acids to fatty alcohols is not popular in industrial productions, since an even higher temperature is necessary and consequently larger amounts of by-products (hydrocarbons and metal soaps)

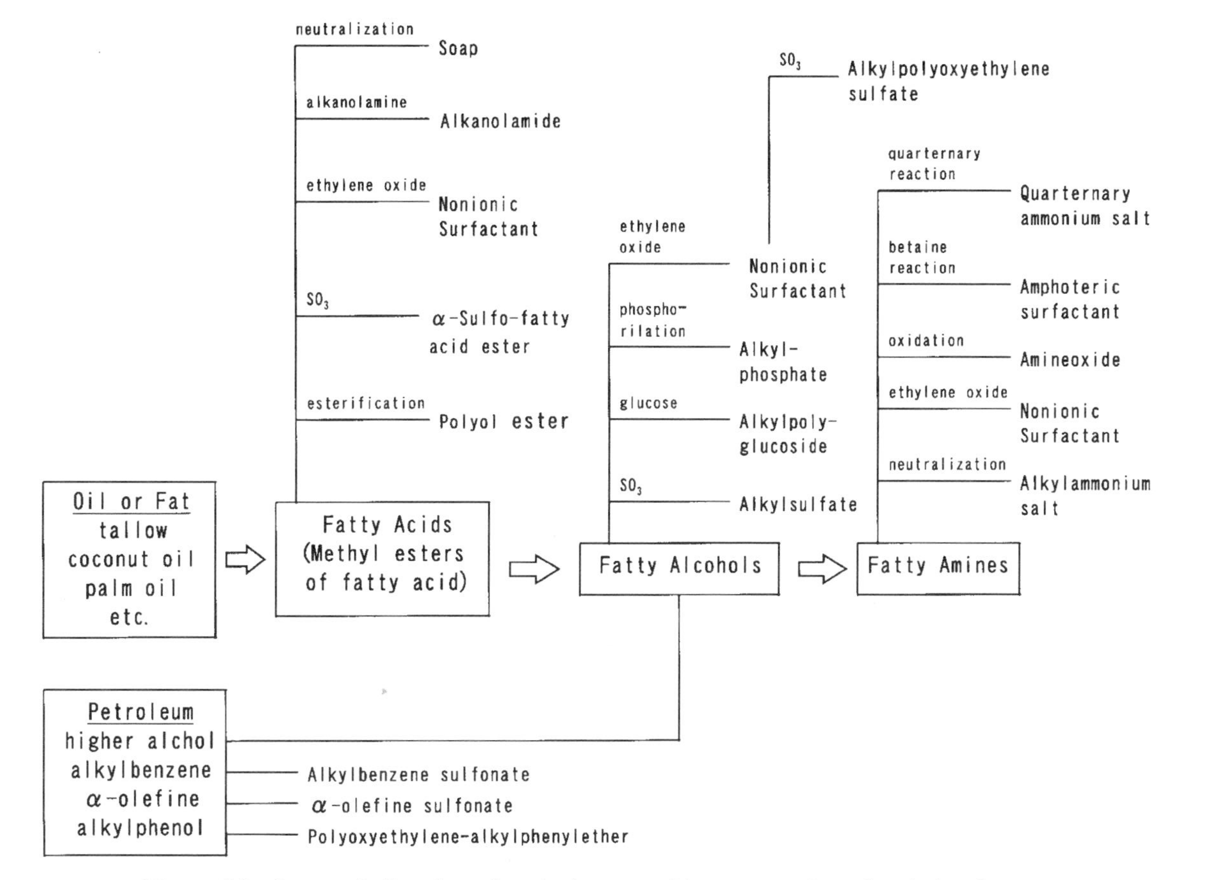

Figure 2.3 Systematic flowchart of synthetic routes of the most popular and typical surfactants.

contaminate the product. Crucially important to reducing the production cost is a new, more efficient catalyst.

c. Synthesis of Fatty Amines from Fatty Acids and/or Alcohols

Fatty amines are nowadays produced by the following three reaction routes.

$$RCOOH \xrightarrow{NH_3} RCN \xrightarrow{2H_2} RCH_2NH_2$$

$$RCN \xrightarrow{H_2} RCH{=}NH \xrightarrow{RCH_2NH_2}$$

$$RCH{=}NCH_2R \xrightarrow{H_2} (RCH_2)_2NH$$

$$ROH \xrightarrow{HCl} RCl \xrightarrow{HN(CH_3)_2} RN(CH_3)_2$$

$$ROH \xrightarrow{HN(CH_3)_2} RN(CH_3)_2$$

Primary amines are produced through nitrils. Nitrils are synthesized from fatty acids and ammonia with an alumina catalyst at 300–350°C. The nitrils are then reduced to primary amines with a Raney nickel catalyst at 130–150°C under the hydrogen pressure of 10–30 atm. Dialkylamines (secondary amines) are also obtained by the reduction of nitrils utilizing the Raney nickel catalyst at about 200°C. Tertiary amines can be produced by two kinds of reaction. One is the reaction of alkylchloride and dimethylamine, which is done at 110–150°C and the pressure of 5–10 atm. The other is the direct amidation reaction of fatty alcohol, where fatty alcohols are reacted with dimethylamine at 150–250°C by the catalysis of copper-nickel or copper-chromium.

d. Derivation to Various Kinds of Surfactant

Fatty acids, alcohols, and amines are derived to anionic, cationic, amphoteric, and nonionic surfactants by a variety of reactions—neutralization, sulphation, sulfonation, esterification with phosphoric acid, ethoxylation, amidation, carboxymethylation, oxidation and quarternary ammonium reactions, and so on. The same types of surfactant can be, of course, derived from petroleum raw materials with the same functional groups (–COOH, –OH and $-NH_2$).

2.2.2 *ANIONIC SURFACTANTS*

Anionic surfactants are electrolytes, and a surface active ion is an anion when the surfactants dissociate in water. The anionic surfactants adsorb on various kinds of substrates and give them an anionic charge. This action of the anionic surfactant contributes to the strong detergency and high foaming power of the agent. Consequently, anionic surfactants are now being used most widely and extensively in detergents, shampoos, body cleansers, and so on.

a. Soap (Sodium Salt of Fatty Acid)

Chemical structure : $RCOO^- Na^+$
Soap is the oldest known surfactant and has been used since the Egyptian era. It is used mostly as toiletry soap bars for body cleansers and sometimes for fabric detergents. In detergents, soap is sometimes formulated as a foam-control agent. Direct saponification from fats was used in old-time soap industry, but today soap is obtained by just neutralizing fatty acids with sodium hydroxide. Soap is a salt of weak acid (fatty acid) and strong base (sodium hydroxide), and shows an alkaline nature (pH~10) in aqueous solutions. This nature favors the detergency, but can cause skin irritation and hair damage. In addition, soap is not tolerant of hard water and forms scum with calcium or magnesium ions.

b. Alkyl Sulfate (AS)

Chemical structure: $ROSO_3^- M^+$
Sodium dodecyl sulfate (SDS) is the most typical AS compound, and is used extensively in scientific research. Alkyl sulfates are synthesized from higher (fatty) alcohols by a sulphation reaction with SO_3 gas, chlorosulfonic acid ($ClSO_3H$), or concentrated sulphuric acid. In industrial production, higher alcohol is reacted with SO_3 gas as a thin liquid film. AS is widely used as an emulsifier, a dispersing agent for toiletries and pharmaceuticals, a foaming agent, cleanser etc. Dodecyl sulfate in particular is an essential component of shampoos and a foaming agent for toothpaste. AS is also employed as the main component of detergent for wool; a wetting agent for pesticide-, insecticide-, and weeding-powders; an anticaking agent for fertilizers; and so on. Triethanolammonium salt of AS is often used when low Krafft point (the temperature below which surfactant crystal precipitates out, see Section 3.3.1) is necessary in practical applications.

c. Alkylpolyoxyethylene Sulfate (AES or ES)

Chemical structure: $RO(CH_2CH_2O)_mSO_3^- M^+$
AES is an improved surfactant of AS: it has greater solution stability at low temperatures, has more tolerance for hard water, and causes fewer skin irritations. This surfactant is one of the rare anionic agents that is soluble in water in the presence of calcium and/or magnesium ions. Consequently, AES is used in applications similar to AS, as well as being a main component in heavy-duty (fabric) and dishwashing detergents. One weak point of this surfactant is the caking phenomenon that results from its hygroscopic nature. Thus, some special technology for anticaking is needed for AES to be used in powder detergents. AES is synthesized from alkylpolyoxyethylene ether (one kind of nonionic surfactant) by the sulphation reaction mentioned above.

d. Alkylbenzene Sulfonate (LAS, ABS)

Chemical structure:

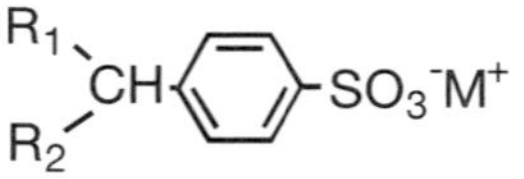

$$(R_1)(R_2)CH-C_6H_4-SO_3^-M^+$$

Alkylbenzene sulfonate is one of the most extensively used anionic surfactants as the main component of fabric, dishwashing, and industrial-use detergents. It is not used in shampoos because it causes hair to feel stiff after washing. The agent is also applicable as an emulsifier and a dispersing agent for agricultural chemicals. Alkylbenzene produced from a petroleum source is sulfonated with SO_3 gas or fuming sufuric acid to convert to the LAS or ABS. A benzenesulfonate group is randomly attached to an alkyl chain in its position.

Linear alkylbenzene sulfonate (LAS), having a linear alkyl chain, is easily biodegraded and thus is called a soft-type agent. On the other hand, ABS, having a branched chain, is not easily biodegraded and thus is called hard-type alkylbenzene sulfonate. Hard-type ABS used to cause the foaming problems in rivers flowing in big cities, but the problems were solved by employing LAS. Soft-type alkylbenzene sulfonate, i.e., LAS, is now mainly used in most of the developed countries.

This surfactant has a poor tolerance for hard water because of the high Krafft point of calcium and/or magnesium salts, and requires some sequestering agents for these divalent ions to be applied to a fabric detergent (Tsujii *et al.*, 1980). Sodium polyphosphates were used as the chelating (sequestering) agents until it was discovered that polyphosphates are one of the causes of eutrophication in lakes. Now different types of sequestering agents such as zeolites are used (see Section 4.2.1.*d*).

e. α-Olefin Sulfonates (AOS)

Chemical structure: Mixture of $RCH{=}CHCH_2{-}SO_3^- M^+$ and $RCH(OH)CH_2CH_2SO_3^- M^+$

The first part of the mixture is the alkenyl type and the second part the hydroxyl type. AOS is obtained from α-olefin by reaction with SO_3 gas. Because its tolerance for hard water and its tendency for caking are intermediate between LAS and AES, one can say that AOS is a well-balanced or noncharacteristic surfactant. AOS is one of the main components for fabric and dishwashing detergents.

f. Monoalkyl Phosphate (MAP)

Chemical structure: $ROP({=}O)(OH)(O^- M^+)$

Fatty alcohol is esterified with phosphoric acid to give monoalkyl phosphate. Phosphorus oxychloride and phosphorus pentoxide are also employed for this esterification reaction. MAP is a dibasic acid, so we can have mono-neutralized and di-neutralized material. Special care is taken in synthesizing pure monoalkyl phosphate to avoid a by-product of dialkyl phosphate that depresses the foaming property of MAP. The most favorable characteristic of this surfactant is low skin irritation. Unlike soaps, MAP works in neutral pH and thus is widely used as face cleansers and body shampoos, especially in Japan.

g. Acyl Isethionate

Chemical structure: $RCOOCH_2CH_2SO_3^- M^+$

This surfactant causes only mild skin irritation and is used as a body cleanser. Soap bars mixed with acyl isethionate are popular particularly in the United States and Europe. Acyl isethionate disperses soap scum and also changes the washing- and rinsing-feeling of the soap.

h. Acyl Glutamate

Chemical structure: $RCONHCH(CH_2CH_2COOH)COO^- M^+$

This is one of the typical surfactants derived from amino acids. Acyl glutamate is sometimes used as a face or body cleanser because it causes only mild skin irritation.

i. *N*-Acyl Sarcosinate

Chemical structure: $RCON(CH_3)CH_2COO^- M^+$
This long-known surfactant is an amino acid surfactant. It is sometimes used as a foaming agent of toothpaste as well as a cosurfactant of shampoo.

j. Alkenyl Succinate

Chemical structure: $RCH{=}CHCH_2CH(COOH)CH_2COO^- M^+$
This surfactant is synthesized from α-olefin and maleic anhydride. It is sometimes used as a surfactant for emulsion polymerization and a cosurfactant of dishwashing detergent.

2.2.3 CATIONIC SURFACTANTS

Cationic surfactants are also electrolytes, with a positive charge. Most materials have negative charges in an aqueous media, and the cationic surfactant molecules adsorb by orienting their hydrophilic head group toward the surface of the materials. This characteristic nature of the cationic surfactant is fully utilized in several products, such as fabric softeners and hair conditioners. Some cationic surfactants can be also used as a germicide, probably because of their strong ability to adsorb to negatively charged materials in the living bodies of bacteria such as chondriosomes, proteins, and nucleic acids. All the cationic surfactants used at present are derivatives of alkylamines. The synthetic routes of alkylamines were described in Section 2.2.1.*c*.

a. Alkyltrimethylammonium Salts

Chemical structure: $RN^+(CH_3)_3X^-$
This surfactant is the most typical cationic surfactant. Tertiary amine, $RN(CH_3)_2$, is reacted with methylchloride or methylbromide to be converted to quaternary ammonium salt. Methylchloride is commonly used in industrial productions, so X^- is usually a chloride ion. This agent is applied to hair conditioners and antistatic agents.

b. Dialkyldimethylammonium Salt

Chemical structure: $R_2N^+(CH_3)_2X^-$
Dialkyldimethylammonium salt is synthesized by the reaction of dialkylamine with methylchloride. This surfactant has two long hydrocarbon chains in one

molecule, and is well known as the first synthetic surfactant found to form vesicles (Kunitake and Okahata, 1977; Deguchi and Mino, 1978; Tran *et al.*, 1978). The agent is mainly used for fabric softeners, but is not easily biodegraded. More biodegradable cationic surfactants targeted for the softeners are now being developed extensively around the world (see Section 4.3.4.*a*).

c. Alkylammonium Salt

Chemical structure: $RNH_3^+X^-$

Neutralization of primary amine with hydrochloric acid or acetic acid gives this surfactant. Rust preventive agents for metals, mineral flotation reagents, and emulsifiers for asphalt are the main application of this surfactant.

d. Alkylbenzyldimethylammonium Salt (Benzalkonium salt)

Chemical structure:

$$RN^+\text{—}(CH_2C_6H_5)(CH_3)_2X^-$$

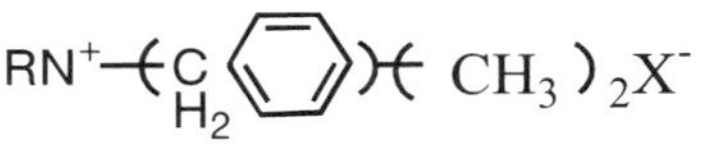

This cationic surfactant is popular as a germicide or a sterilizer. It is often called "invert soap" because of its opposite ionic charge from soap. The reaction of tertiary amine $RN(CH_3)_2$ with benzylchloride is used to produce this surfactant.

e. Alkylpyridinium Salt

Chemical structure:

$$RN^+C_5H_5\ X^-$$

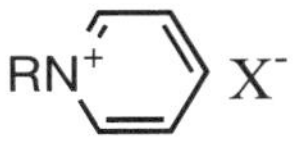

Alkylhalide is reacted with pyridine to synthesize this material. This surfactant is used mainly for pure scientific research and very rarely in germicides or bactericides.

2.2.4 AMPHOTERIC SURFACTANTS

Amphoteric surfactants are capable of variation in their ionic nature with the changing pH of solutions. Thus, this surfactant can be in the anionic, nonionic, cationic, or zwitter-ionic state depending on the pH value of the solution it is in. Amphoteric surfactants are hardly ever used as a main component of any product. Their primary use is as an important cosurfactant that boosts the detergency and the foaming power of anionic surfactants. This boosting effect is known to be due to the *addition compound* formation with anionic surfactants in their aqueous solutions (Tajima *et al.*, 1979; Tsujii *et al.*, 1982, see Section 4.1.4).

a. Alkyldimethylamine Oxide

Chemical structure: $RN(CH_3)_2O$
This compound is a typical booster for anionic surfactants such as alkylsulfate and alkylpolyoxyethylene sulfate, mainly used in dishwashing detergents. The amine oxide is synthesized from tertiary amine, $RN(CH_3)_2$, by oxidation reaction with hydrogen peroxide. The varying ionic nature of this surfactant shown here depends on the pH values of the solutions.

$$RN(CH_3)_2O \underset{-H^{\oplus}}{\overset{+H^{\oplus}}{\rightleftharpoons}} \overset{\oplus}{R}N(CH_3)_2OH$$

b. Alkylcarboxy Betaine

Chemical structure: $RN^+(CH_3)_2CH_2COO^-$
Monochloroacetic acid is reacted with tertiary amine, $RN(CH_3)_2$, to obtain this amphoteric surfactant. This surfactant is used as a booster in dishwashing detergents and shampoos. Shown here is the pH-dependent ionic nature of this compound.

$$\overset{\oplus}{R}N(CH_3)_2CH_2CO\overset{\ominus}{O} \underset{-H^{\oplus}}{\overset{+H^{\oplus}}{\rightleftharpoons}} \overset{\oplus}{R}N(CH_3)_2CH_2COOH$$

c. Alkylsulfobetaine

Chemical structure: $RN^+(CH_3)_2CH_2CH_2CH_2SO_3^-$ or
$RN^+(CH_3)_2CH_2CH(OH)\ CH_2SO_3^-$
The first compound shown is synthesized by the reaction of tertiary amine with propane sultone, the latter by the reaction with epichlorohydrin and then sulfonation. This sulfobetaine type surfactant shows a boosting effect similar to the pre-

vious two amphoteric surfactants as well as a unique solution property like the Krafft point depression phenomenon on the addition of inorganic salts (Tsujii and Mino, 1978). But the agent is not used practically so far since the propane sultone as a raw material shows a toxicity of mutagenic cancer. The zwitter-ionic form is the normal ionic nature of this surfactant in almost all pH ranges used in experiments since the quarternary ammonium hydroxide is a strong base group and sulfonic acid is a strong acid group.

d. Amide-Amino Acid Type Surfactant

Chemical structure: $RCONHCH_2CH_2N^-$ $(CH_2CH_2OH)(CH_2COO^- M^+)$
This surfactant is sometimes used as a main component of shampoos because of its low skin irritation nature. It is a rare case for amphoteric surfactant to be used as a main ingredient of any products. Alkylimidazoline is hydrolyzed to amide-amine, and then the amide-amine is carboxymethylated to convert to this surfactant.

e. Phosphatidyl Choline (Lecithin)

See Section 2.3.

2.2.5 NONIONIC SURFACTANTS

Nonionic surfactants are not electrolytes, and have some nondissociative hydrophilic group. Polyoxyethylene groups, saccharides (sorbitan, sugar, glucose, etc.), and hydroxyl groups are examples of nonionic hydrophilic groups. Nonionic surfactants are mostly tolerant in aqueous solutions of added salts and/or hard water.

a. Polyoxyethylene Alkyl Ether

Chemical structure: $RO(CH_2CH_2O)_mH$
This nonionic surfactant is one of the most popular emulsifiers, widely used in the emulsions of toiletries, cosmetics, pharmaceuticals, fungicides, insecticides, and herbicides. One can control the hydrophile/lipophile balance (HLB, see Section 3.2.3.*c*) of the surfactant by changing the polyoxyethylene chain length. The agent bearing longer polyoxyethylene chain is more hydrophilic and has a larger HLB number. Its detergency for oily dirt is good, but this surfactant has not been used for a long time in powder detergents for fabrics because of caking prob-

lems: It was not easy to make a good powder detergent using a surfactant in a liquid state. An inorganic material developed recently overcomes this problem and makes this nonionic surfactant applicable to fabric detergents. This surfactant is produced by addition polymerization of ethylene oxide to higher alcohol with the aid of alkaline or acid catalyst. The acid catalyst (such as tin tetrachloride) gives a narrower molecular weight distribution to the products but also produces more by-product such as dioxane.

b. Polyoxyethylene Alkylphenyl Ether

Chemical structure:

$$R-C_6H_4-O(CH_2CH_2O)_mH$$

This surfactant is used for almost the same purpose as polyoxyethylene alkyl ether. The branched alkyl chain type of this agent is not biodegraded well, so its applications are limited. Addition of ethylene oxide to alkylphenol gives this compound.

c. Polyoxyethylene Fatty Acid Ester

Chemical structure: $RCOO(CH_2CH_2O)_mH$
This is also an emulsifier showing properties similar to the above two surfactants. This compound is an ester (not ether) produced by addition of ethylene oxide to fatty acid. It can be hydrolyzed, a property that is sometimes of merit and sometimes not in practical applications.

d. Sorbitan Fatty Acid Ester

Chemical structure:

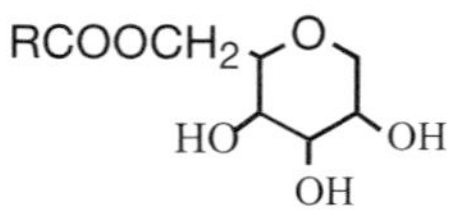

RCOOCH$_2$ O
HO OH
OH

This nonionic surfactant is very popular under the commercial brand name SPAN (ICI, America) and is a most typical emulsifier. SPAN is commonly a hydrophobic surfactant usually used in combination with the more hydrophilic surfactant TWEEN (2.2.5.*e*). The HLB system used to select the most suitable emulsifiers was first invented to utilize SPAN and TWEEN by the supplier of these surfactants. This surfactant is synthesized by esterification of sorbitan with fatty acid and is used as a food additive.

e. Polyoxyethylene Sorbitan Fatty Acid Ester

Chemical structure:

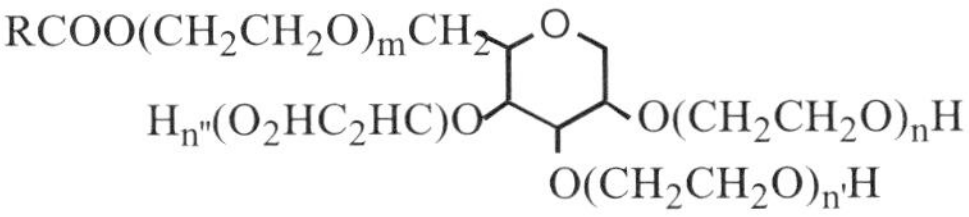

$RCOO(CH_2CH_2O)_mCH_2$ O

$H_{n''}(O_2HC_2HC)O$ $O(CH_2CH_2O)_nH$

$O(CH_2CH_2O)_{n'}H$

This surfactant is known under the commercial brand name TWEEN. TWEEN is produced by the addition of ethylene oxide to SPAN and is more hydrophilic than SPAN because of the added polyoxyethylene chains. This emulsifier is a food additive in the United States, but not in Japan.

f. Sugar Ester of Fatty Acid

Chemical structure:

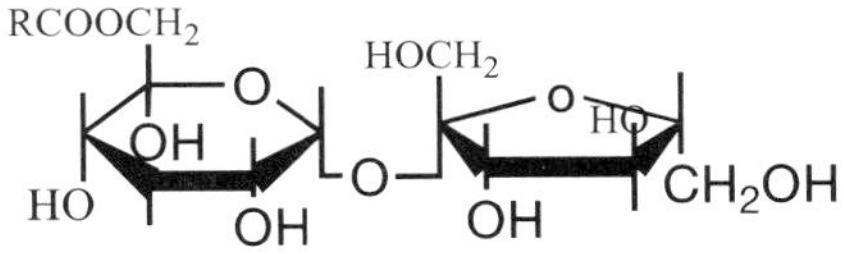

$RCOOCH_2$ $HOCH_2$

O OH O O HO

HO OH OH CH_2OH

This surfactant is frequently used as a relatively hydrophilic food additive and is useful particularly in Japan since TWEEN is not permitted there by law. This agent is mainly used as an emulsifier for foods.

g. Alkyl Polyglucoside

Chemical structure:

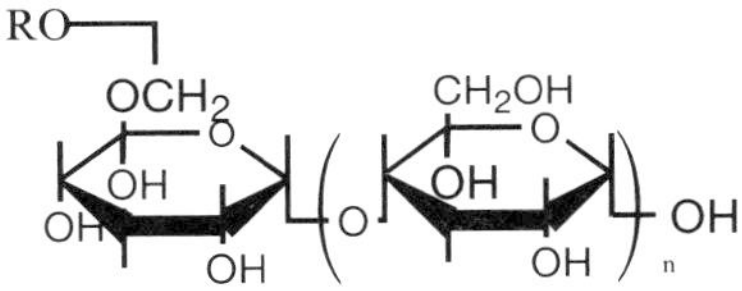

RO

OCH_2 CH_2OH

O OH O OH O

OH OH OH OH

n

Fatty alcohol is reacted with glucose by an acid catalyst such as *p*-toluene sulfonic acid to produce alkyl polyglucoside. This surfactant is mild to the skin and is used in dishwashing detergents as well as shampoos. The number of glucose units in one molecule is usually 1–2, i.e., $n = 0–1$.

h. Fatty Acid Diethanolamide

Chemical structure: $RCON(CH_2CH_2OH)_2$
This nonionic surfactant is mainly used as a booster for detergency and foaming power in shampoos and dishwashing detergents. This boosting effect is similar to that of amphoteric surfactants (such as amine oxide and/or carboxybetaine-type surfactants) but the *addition compound* formation between this surfactant and anionic ones is not known. Fatty acid chloride is reacted with diethanolamine to obtain this compound.

i. Fatty Acid Monoglyceride

Chemical structure: $RCOOCH_2CH(OH)CH_2OH$
This surfactant is used as an emulsifier for foods. Pure monoglyceride is hardly ever obtained; it is usually contaminated with a considerable amount of diglyceride, etc. This surfactant has a low HLB number and is not soluble in water. Some of the agent swells in water and forms liquid crystals. Polyglycerine esters of fatty acid are also available as more hydrophilic derivatives.

j. Alkylmonoglyceryl Ether

Chemical structure: $ROCH_2CH(OH)CH_2OH$
This is an ether-type surfactant of the above fatty acid monoglyceride. This surfactant is also not soluble in water and forms liquid crystal. Liquid crystal emulsification of this nonionic surfactant gives a very interesting W/O emulsion (see Section 4.4.2.*a*). This emulsion can contain more than 80–90 wt % water, but is still a water-in-oil-type emulsion! This characteristic emulsifying property has been applied to cosmetic foundations.

k. Fatty Acid Propyleneglycol Ester

Chemical structure: $RCOOCH(CH_3)CH_2OH$
This is one of the emulsifiers for food that has a low HLB number. Compared with fatty acid monoglyceride, this compound has a lower melting point and can be easily handled.

2.2.6 SURFACE ACTIVE OLIGOMERS AND POLYMERS

There are several kinds of oligomers and polymers that are used like a surfactant. They do not have the molecular structure of a typical surfactant (the hydrophilic head and hydrophobic tail), but can still be used, in particular, as dispersing and/or coagulating agents. Polymeric dispersants of this type, called *protective colloids,* have been known for a long time. The stabilizing mechanism of dispersions by these polymers and oligomers recently has been made clear and will be discussed later in detail (Section 3.2.4.*c*). In addition, the water-soluble polymers, particularly cationic ones, are used as coagulating agents. Surface active oligomers and polymers are listed in Figure 2.4.

a. Formaldehyde Condensates of Naphthalene Sulfonate

This is one of the most typical oligomer-type dispersants, most popularly used as a cement dispersing agent. This oligomeric dispersant is also applied to pigments, agrochemicals, rubber, and latex dispersions. Its low foaming property is a good characteristic of this dispersant. Oligomers having 3–5 naphthalene nuclei are suitable for use as a general dispersing agent and those with more than 10 nuclei are particularly useful for cement dispersants. Poorly weather-resistant, the agent is not suitable for emulsion paints and paper coatings. Naphthalene sulfonate synthesized by sulfonation reaction of naphthalene with concentrated sulfuric acid is condensed with formaldehyde to produce this agent.

b. Sodium Polyacrylate

This polymer is especially useful in dispersing inorganic powders and pigments. Kaolin, calcium carbonate, aluminum hydroxide, titanium oxide, ferric oxide, etc., are examples that are dispersed well with this agent. This agent has good weather resistance and can be used in applications requiring whiteness and brightness of color, such as paper coatings. It is also utilized in the slurry transportation of inorganic pigments.

c. Copolymer of Olefin and Sodium Maleate

This copolymer is a good dispersant that can be applied to both inorganic and organic powders and pigments. The applications of this agent are emulsion paints, resin paints, disperse dyes, ceramics, agrochemicals, etc. This compound has good weather resistance but a relatively high foaming property. Olefins such as styrene, ethylene, propylene, isobutylene, and octene are copolymerized

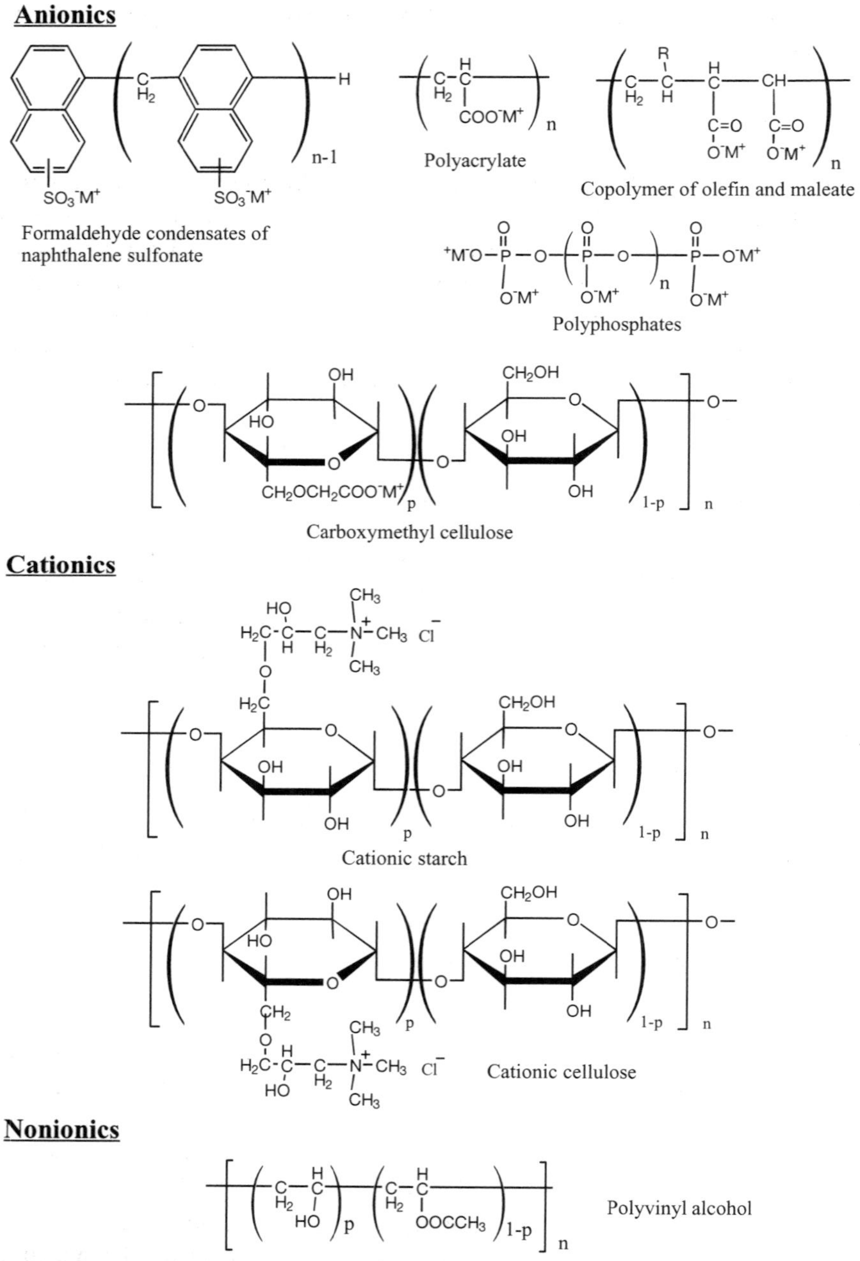

Anionics

Formaldehyde condensates of naphthalene sulfonate

Polyacrylate

Copolymer of olefin and maleate

Polyphosphates

Carboxymethyl cellulose

Cationics

Cationic starch

Cationic cellulose

Nonionics

Polyvinyl alcohol

Figure 2.4 Molecular structures of typical surface active oligomers and polymers.

alternatively with maleic anhydride, and the product polymer is hydrolyzed with sodium hydroxide.

d. Lignin Sulfonate

Lignin sulfonate is obtained in large quantities as a by-product when pulp is produced from wood chip. This compound is a very complicated polymeric material that consists of phenyl-propane networks having a sulfonate group mainly attached to the α-position of a side chain. The product is a mixture of many kinds of structures and has a molecular weight distribution over a wide range—from several hundred to several million. This dispersing agent is cheap and used widely in the fields of cement, ceramics, disperse dyes, etc. Its strong dark brown color somewhat limits its use.

e. Polyphosphate

This is an inorganic and low-priced dispersing agent, particularly useful for inorganic powders. This compound has a weak point of hydrolysis at high temperatures. Sodium tripolyphosphate was widely used for detergents until the 1970s; its use is now very limited because of its contribution to the eutrophication of lakes.

f. Carboxymethyl Cellulose

Cellulose is reacted with monochloroacetic acid and converted to this water-soluble anionic polymer. This polymer is mainly used in detergents as an antiredeposition agent for soils; it is also used in pigment dispersions.

g. Cationic Cellulose

A quarternary ammonium ion is attached to the –OH group of cellulose with glycidyltrimethylammonium chloride to synthesize the cationic cellulose. Cationic cellulose is water-soluble and used as a coagulating agent in a variety of applications. The complex-formation phenomenon of the polymer with anionic surfactants is useful for obtaining the conditioning effect of shampoos (see Section 4.2.3.*a*). The complex is water-insoluble but is solubilized by the excess amounts of anionic surfactant in shampoo formulas. When the shampoo is diluted on washing and/or rinsing, the complex is precipitated out onto hair fiber and gives it a lubrication effect.

h. Cationic Starch

Cationic starch is produced in the same way as cationic cellulose. This cationic polymer is used as both a dispersing and a coagulating agent, depending on its molecular weight, the number of attached cationic groups in a polymer chain, and the conditions of its applications. It is often used as a dispersing agent for emulsion polymerization, but it can also be employed as a hetero-coagulation agent when used as a fixer for inorganic fillers and/or sizing agents in the papermaking industry.

i. Polyvinyl Alcohol

This nonionic polymer is frequently used as a dispersing agent in emulsion and/or suspension polymerization, particularly in emulsion polymerization of vinyl acetate. Polyvinyl alcohol is obtained by the hydrolysis of polyvinyl acetate and shows best solubility in water when the degree of saponification is ca. 80%.

j. Polyethylene Glycol

This polymer is sometimes used as a dispersing agent for both fabric and dishwashing detergents.

k. Polyacrylamide

High-molecular-weight (~million) products of this nonionic polymer are sometimes used as coagulating agents, especially as fixing agents of fillers and/or sizing agents in a papermaking process.

2.3 Naturally Produced Surface Active Compounds

A biological body consists of 70% water and many water-insoluble materials. So one can easily imagine that surface active substances play some important roles in biological cells in order to maintain the biological activities of living systems. For example, some microorganisms are well known to assimilate water-insoluble hydrocarbons: They release certain surface active substances to emulsify the hydrocarbon oils and take the hydrocarbons into their cells. Surface active substances working in and produced from natural living systems are listed in Figure 2.5.

Phospholipid

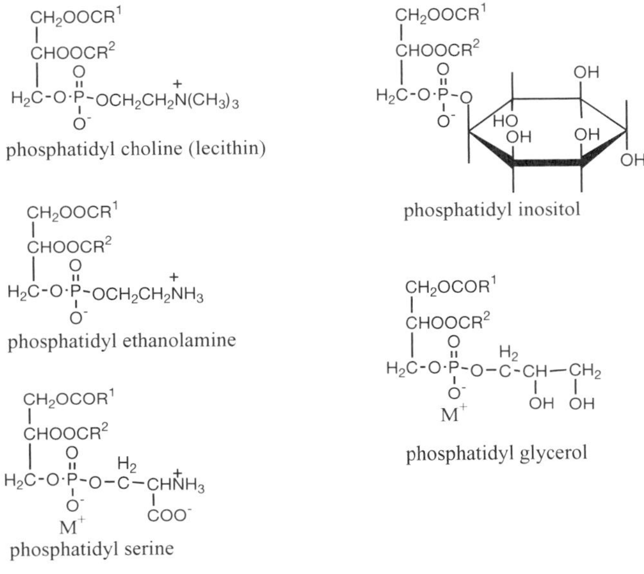

Glycolipid

Sophorolipid

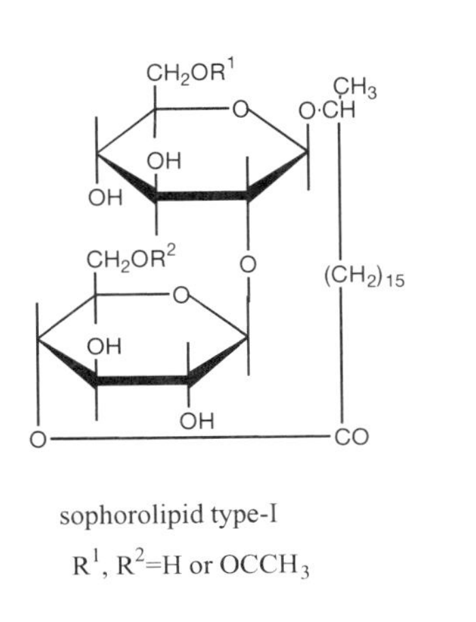

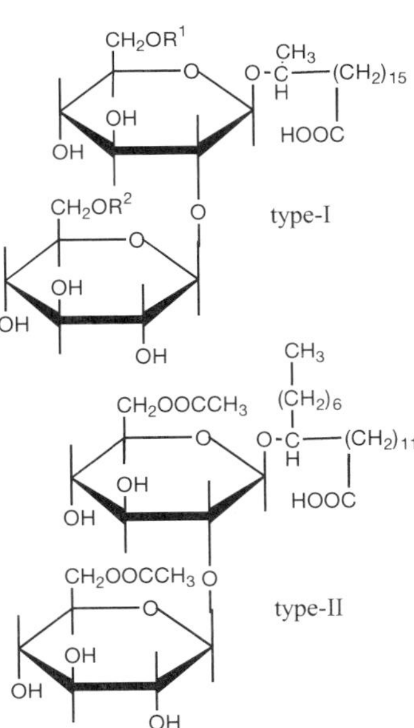

Figure 2.5 Some typical examples of naturally produced surface active substances.

Glycolipid (continued)

Rhamnolipid

type-II

rhamnolipid type-I

type-IV

type-III

Trehaloselipid

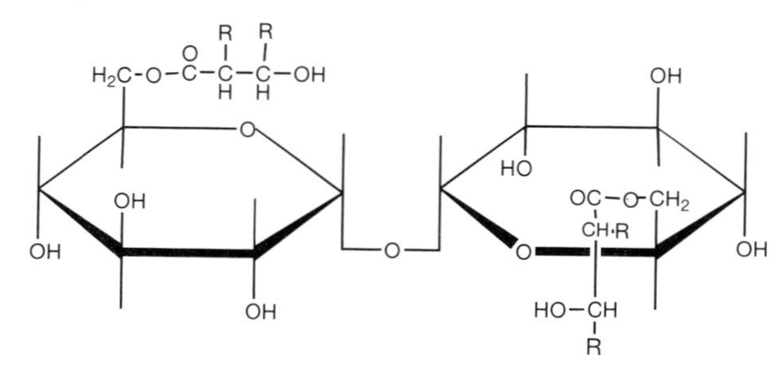

Figure 2.5 *continued.*

Bile acid (main components)

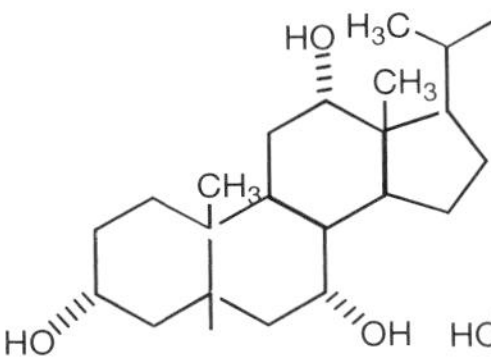

cholic acid

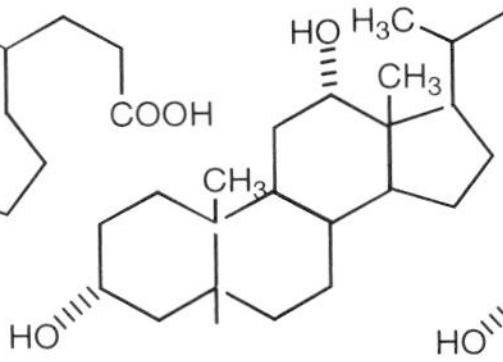

deoxycholic acid

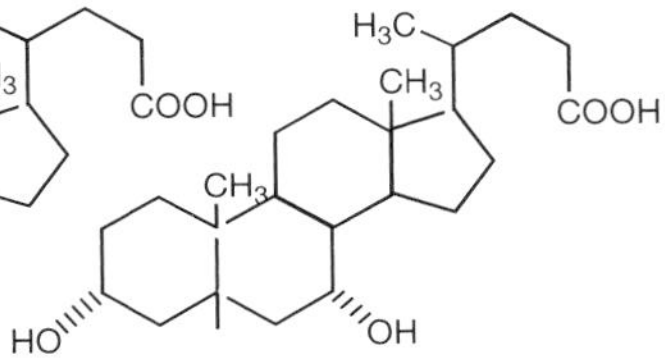

chenodeoxycholic acid

Rosin (main components)

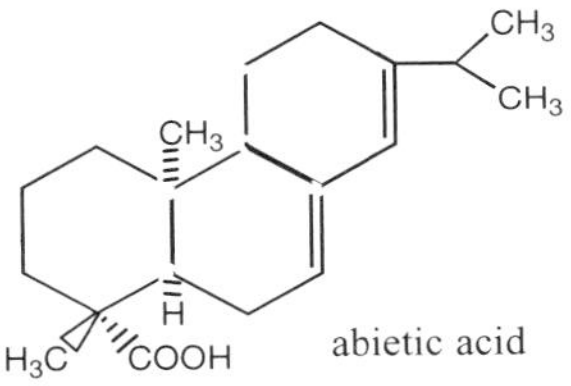

abietic acid

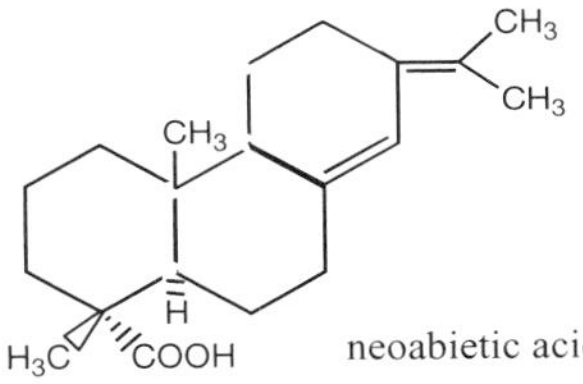

neoabietic acid

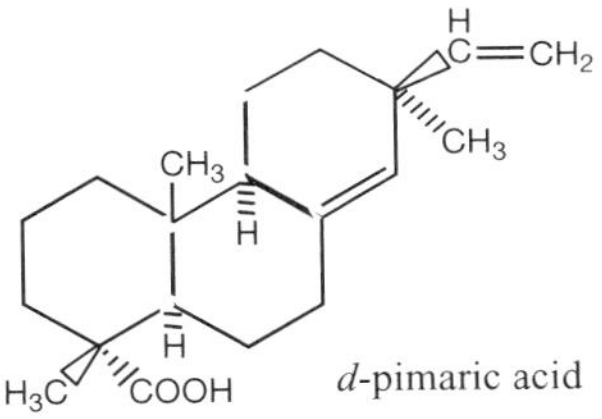

d-pimaric acid

Cholesterol

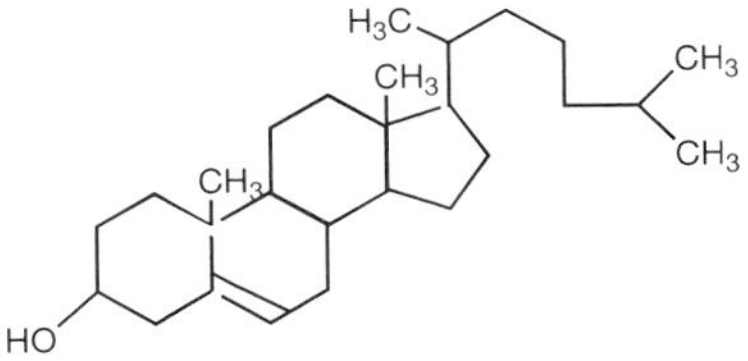

Peptide derivative

surfactin

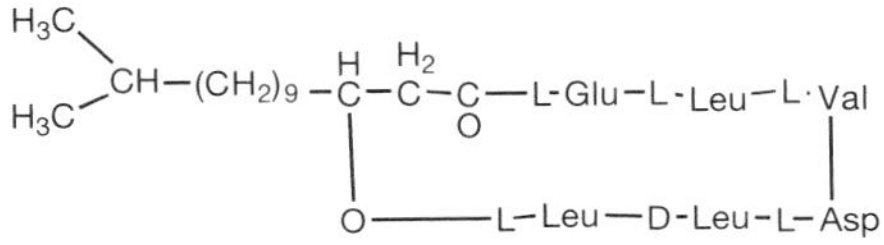

Figure 2.5 *continued.*

2.3.1 PHOSPHOLIPIDS

Phospholipids are of special importance as they are the main component of biological membranes. Biological membranes are, of course, constructed of many other components as well, such as proteins and glycolipids. But the fundamental bilayer structure of the membranes is formed by phospholipids, and one can regard them as the essential material for biological membranes (see Section 5.2.1).

There are two kinds of phospholipid: glycero-phospholipid and sphingo-phospholipid. Glycero-phospholipids are esters of glycerin with fatty acids and phosphoric acid or its derivatives. Esterification of the α- and β-positions of glycerin with fatty acid gives the double-chain type of phospholipid.

The lyso-type phospholipid is a single-chain molecule and has a free —OH group in the β-position. In aqueous solutions, the double-chain lipids readily form liposomes that exhibit most of the fundamental properties of biological membranes (see Section 3.3.4). Single-chain lyso-type lipids, on the other hand, form micelles like ordinary water-soluble surfactants. The main components of biomembranes are, of course, the double-chain-type phospholipids.

Table 2.1 shows some examples of phospholipid composition in biological membranes (rat lever cells) (Longmuir, 1987). Most of the lipids are the double-chain type. Phosphatidyl choline (lecithin), the most typical double-chain lipid, is used frequently in the research of bilayer membranes, liposomes, and so on. Phosphatidyl ethanolamine, phosphatidyl serine, and phosphatidyl inositol are also sometimes used in membrane mimetic research. Soybean lecithin is produced industrially and utilized mainly as a food additive.

The enzymatic modification technology of phospholipids is worth noticing. Transphosphatidylation reaction by phospholipase D makes it possible to exchange the polar group of a phospholipid. The content of the particular structure of phospholipid can be increased, and furthermore a new type of polar group can be inserted by this enzymatic reaction. Modified lecithins by phospholipase A_2 and/or D are used in food industries.

2.3.2 GLYCOLIPIDS

Polysaccharide chains located on the surface of biological membranes are known to play an important role in enabling living cells to recognize each other. The determining factor of blood type and signals for contact inhibition between multiplying cells are such examples of this cell–cell recognition. These polysaccharide chains are, in many cases, present in the form of glycolipids. However, the lipids have not been studied yet as a surface active substance because there are too few

Table 2.1 Phospholipid Composition (%) in Biological Membranes of Rat Lever Cells

Phospholipid	*Nucleus*	*Endoplasmic Reticulum*	*Golgi Apparatus*	*Mitochondria*		*Lysosome*	*Plasma Membrane*
				Inner	*Outer*		
Phosphatidyl choline	61.4	60.9	45.3	45.4	49.7	41.6	34.9
Sphingomyeline	3.2	3.7	12.3	2.5	5.0	9.1	17.7
Phosphatidyl ethanolamine	22.7	18.6	17.9	25.3	23.2	27.3	18.5
Phosphatidyl serine	3.6	3.3	4.2	0.9	2.2	—	9.0
Phosphatidyl inositol	8.6	8.9	8.7	5.9	12.6	9.4	7.3
Phosphatidyl glycerol	—	—	—	2.1	2.5	—	4.8
Cardiolipin	—	—	—	17.4	3.4	—	—
Phosphatidic acid	1	—	—	0.7	1.3	—	4.4
Lysophosphatidyl choline	1.5	4.7	5.9	—	—	1.9	3.3
Lysophosphatidyl ethanolamine	—	—	6.3	—	—	1.3	—
Bis(mono-acylglycerol) phosphate	—	—	—	—	—	4.0	—

of them to be investigated in bulk solutions. All glycolipids listed in Figure 2.5 as surface active compounds are ones produced by microorganisms.

Rhamnolipids, sophorolipids, and trehaloselipids are used by microorganisms to assimilate *n*-alkanes. One microorganism, *Pseudomonas aeruginosa,* produces and releases rhamnolipids to emulsify *n*-alkanes and transfer them into its body. It is quite interesting to note that the rhamnolipids are species-specific: they are active only in the case of assimilation of *Pseudomonas aeruginosa.* Sophorolipids and trehaloselipids are not so specific for the species of microorganisms.

2.3.3 PROTEINS AND PEPTIDE DERIVATIVES

Some proteins exhibit surface activity; casein, for instance, works as an emulsifier of fats in milk. Thus, casein is used in food industry as an emulsifier. We also know that protein solutions are often foaming. Furthermore, most of the membrane proteins are amphiphilic. They have a hydrophobic part to be in contact with membrane lipid, and a hydrophilic portion to be in contact with the aqueous medium. Some of them—say, cytochrome P-450—form an aggregate in water like the micelle formation of a surfactant.

A surface active peptide derivative called *surfactin* is produced by a microorganism of *Bacillus subtilis.* Surfactin is an amphiphile with a hydrophobic hydrocarbon chain and a hydrophilic cyclic peptide. This compound was reported to be more surface active than a typical anionic surfactant (sodium dodecyl sulfate) and was extensively studied as a thrombolytic agent at one time. However, it is not practical to use because of its low productivity and toxic problem. Melittin, a bee poison, is an amphiphilic peptide. Melittin is believed to be poisonous for biomembranes because of its amphiphilic nature (see Section 5.3.1.*b*). Acyl amino acids and/or acyl peptides are also found in natural materials, but in small amounts. They are chemically synthesized and used in practical applications.

2.3.4 OTHER SURFACE ACTIVE NATURAL SUBSTANCES

a. Cholesterol

Cholesterol itself is insoluble and even not dispersive in water. It works, however, when blended in another amphiphile. Cholesterol is particularly well known to stabilize the phospholipid bilayer membranes and is used frequently in liposome preparations. This compound is commonly present in almost all animal fats and cell membranes as free molecules or fatty acid esters. Industrial production of cholesterol is made by extraction from cow brain or spinal cord, fish oils, tallow, and so on. Cholesterol was named after cholelithiasis because of the sterol found first in gall stones. It is also called cholesterin.

b. Bile Acid

Bile acid is released from bile and promotes the fat absorption in intestines through emulsification (see Section 5.3.1.*a*). The main components of bile acid are cholic acid, deoxycholic acid, chenodeoxycholic acid, and lithocholic acid. Bile salts are platelike molecules that have hydrophilic groups on one plate side and a hydrophobic nature on the other side. This characteristic molecular structure seems to originate a unique aggregation structure in aqueous solutions, but the decisive conclusion on the structural model has not been given yet. Bile salt is sometimes used in solubilization of membrane proteins as a mild surfactant that will not denature the proteins.

c. Rosin

The main components of rosin are abietic acid, neoabietic acid, *d*-pimaric acid, etc. The most popular application of rosin is as a paper sizing agent. Water-soluble rosin acid salts are precipitated out onto the pulp surface by the addition of aluminum sulfate in the papermaking process, and the hydrophobic precipitate controls the wettability of paper and prevents the ink from blurring. A colloidal dispersion of water-insoluble rosin acids is also commercially available as a paper sizing agent. The esters of rosin acids with methanol and glycerin are viscous liquid, applied to adhesives, varnish, paints, and so on.

d. Saponin

Saponin is a general name for a complicated mixture of saccharine derivatives that are widely present in plants. Aqueous solutions of saponin have a high foaming property. Saponin can be classified into two types: steroid and triterpenoid. Hydrophilic saccharides (glucose, galactose, rhamnose, xylose, pentose, etc.) are attached to hydrophobic steroids and triterpene to give surface active molecules. Saponin has been popularly known for a long time as an agent to stabilize foams, but it is not yet mass produced for this application.

Refreshing Room!

ODD-EVEN NUMBERS OF CARBON ATOMS IN A HYDROCARBON CHAIN

Most of the commercially available surfactants have a hydrocarbon chain with an even number of carbon atoms because they are produced from natural raw materials such as tallow, coconut oil, and palm oil. Natural fats and oils are esters of fatty acids having even number of carbon atoms with glycerol. But why is the number of carbon atoms of the naturally produced fatty acids even? Biosynthesis of fatty acids gives us the answer. The hydrocarbon chain is elongated stepwise by the addition of acetyl-SCoA, which has two carbons in

(continued)

its molecule (see, e.g., Mahler and Cordes, 1971). So we have to chemically synthesize fatty acids to obtain ones with odd carbon numbers.

The melting point of fatty acids exhibits interesting odd-even alterations but that of alcohols does not (Fig. 2.6). The Krafft point of *n*-alkyl sulfates also shows similar alterations (Lange and Schwuger, 1968) because the crystals of these surfactants may be more similar to those of fatty acids than alcohols. The above results also indicate that the Krafft point is a melting point (as mentioned in Chapter 3).

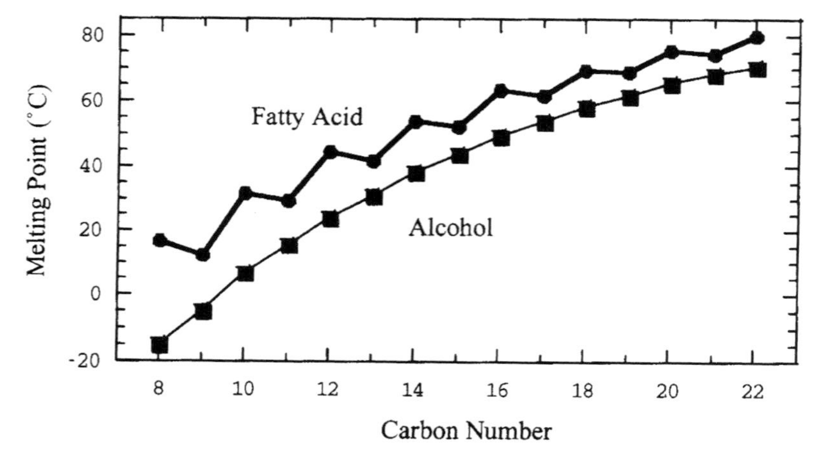

Figure 2.6 Melting points of fatty acids and alcohols with an odd and even number of carbon atoms.

Chapter 3 | Properties and Functions of Surface Activity

3.1 Characteristic Properties of Aqueous Solutions of Surfactants

As we have seen, a surfactant has both hydrophilic and hydrophobic groups in one molecule. All characteristic properties of a surfactant originate from the fact that a hydrophobic group (usually a hydrocarbon chain) is forced to dissolve into water by the hydrophilic group that is attached to the same molecule. The hydrophobic group shows a strong tendency to escape from contact with water molecules and thus likes to adsorb at the surface of the solution and/or at interfaces between the aqueous solution and any hydrophobic solid or liquid particles being present inside the solution. This is why surfactant molecules show very unique (oriented) adsorption. What happens when the surfactant is further added after all such spaces for adsorption are already occupied? The hydrophobic groups then have only one choice: to gather together to avoid contact with the water molecules; in other words, to form aggregates. These two characteristic properties—oriented adsorption and aggregation—result in all kinds of unique properties and functions of the surfactant. Figure 3.1 shows a schematic representation of the fundamental properties of the surfactant described above.

When the hydrophobic group of a surfactant molecule adsorbs at the surface of the solution to avoid contact with water, this adsorption is passive and occurs by a mechanism different from that of ordinary adsorption. In an ordinary case (e.g., adsorption of gases onto solid surfaces), the adsorption occurs due to some attractive interaction between adsorbate and adsorbent. But no such attraction is present between the surfactant molecule's hydrophobic group and the air. Rather, the passive adsorption of surfactant is caused by the strong cohesive energy between water molecules due to their hydrogen bonding (see Section 1.3.1).

Aggregation of surfactant molecules apparently seems due to the attractive interaction between the hydrophobic groups of the agent. However, the attractive interaction energy between hydrocarbons is the same as that between hydrocar-

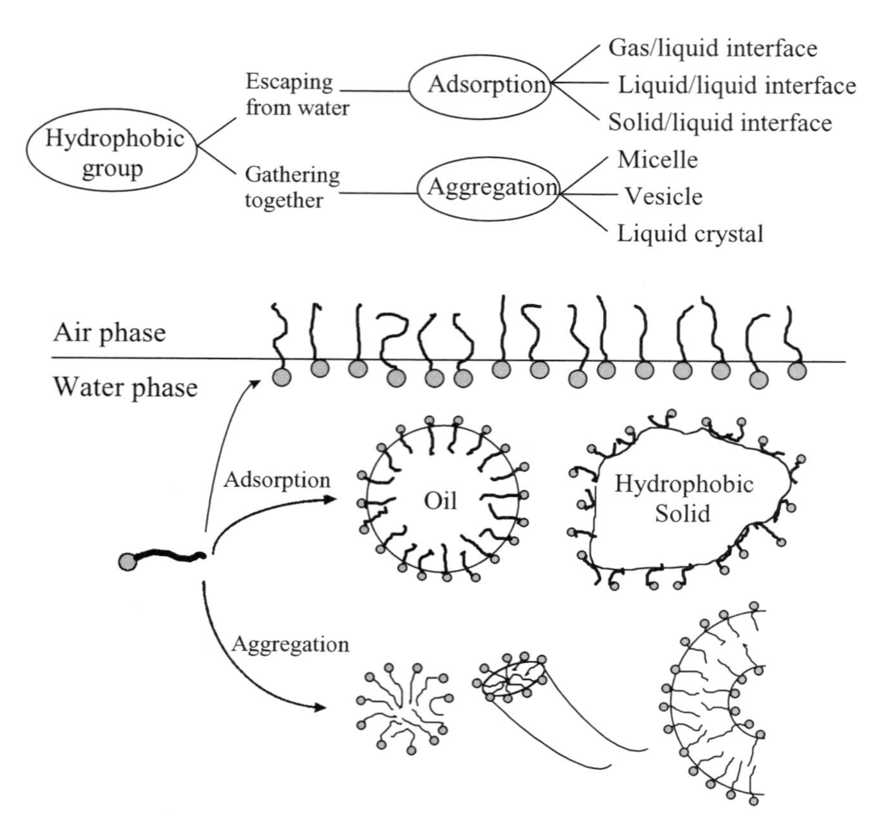

Figure 3.1 Schematic representation of the fundamental properties of the surfactant in aqueous solutions: adsorption and aggregation.

bon and water (Tanford, 1980a). So there is no reason from the viewpoint of the interaction energy as to why the hydrophobic groups of surfactant gather together in aqueous solutions. The aggregation of surfactant molecules also takes place when segregated out of the water phase and is a reflection of hydrophobic interaction as mentioned previously (Section 1.3.1).

3.2 Adsorption and Related Phenomena

As mentioned in the previous section, two fundamental properties of surfactants in aqueous solutions are adsorption and aggregation. In this section, we will describe the adsorption of surfactants and its related phenomena. A surfactant works most efficiently in aqueous solutions and plays an important role in air/water interfaces (surface tension lowering, wettability, foaming, and defoaming), oil/water interfaces (emulsions), and solid/water interfaces (suspensions and coagulations).

3.2.1 SURFACE TENSION LOWERING AND WETTING PHENOMENA

a. Surface Tension Lowering and Adsorption of Surfactants in Aqueous Solutions

The surface tension of water is depressed remarkably with the addition of a surfactant at low (usually less than 1 wt %) concentrations. Indeed, the term *surface active agent* may have come from this characteristic action of the agent. Effective lowering of the surface tension of water by the surfactant means that the surfactant adsorbs strongly at the surface of the aqueous solutions (see Section 1.2 and Eq. (1.5)). When the surfactant molecules adsorb at the surface of an aqueous solution, the water molecules present at the surface till then can move to the bulk solution and make hydrogen bonds fully with surrounding water molecules. The free energy gain of this process is balanced by the surface tension lowering of the solution by the surfactant adsorption (see Section 1.3.1).

Figure 3.2 shows the surface tension-concentration curves of dodecyldimethylammonium chloride in aqueous NaCl solutions (Ozeki *et al.*, 1978). As pointed out in Section 1.2, the adsorption amount of surfactant can be calculated from these γ versus log C curves by applying the Gibbs' adsorption isotherm.

$$\Gamma = -\frac{1}{RT}\frac{d\gamma}{d\ln C} \tag{3.1}$$

This is Eq. (1.5) in Section 1.2. Adsorption amounts, Γ, of surfactant are estimated from the slopes of the curves shown in Figure 3.2. Every curve is convex to the upper side at low surfactant concentrations, i.e., the slope becomes steeper with increasing concentration and attains to a constant value.

Note that the curves suddenly break at certain concentrations and show constant values in surface tension above these concentrations. If the Gibbs' equation is applied to this constant region, the adsorption amount of zero is obtained. But this is not the case in reality. The adsorption amount measured experimentally shows no abnormal behavior at the break point and is, of course, not zero (Muramatsu *et al.*, 1973). It is clear from this result that the application of the Gibbs' adsorption isotherm to the region of constant surface tension in Figure 3.2 is wrong. The breaking in the γ versus log C curve and the leveling off of the sur-

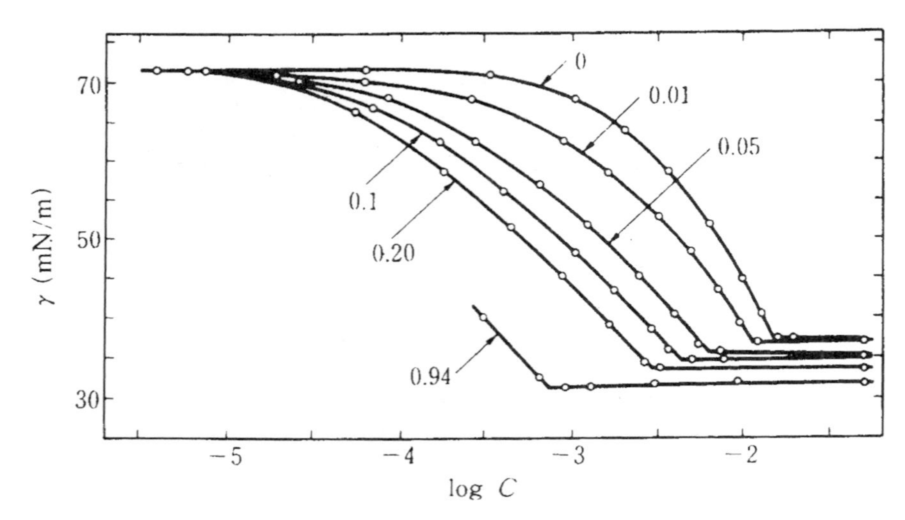

Figure 3.2 Surface tension versus concentration curves of a surfactant (dodecyl-dimethylammonium chloride). The numbers denoted in the figure are NaCl concentrations in mol/l (Ozeki *et al.*, 1978).

face tension are actually due to micelle formation (to be discussed in detail in Section 3.3.2). The break point above which the surface tension becomes constant is called the *critical micelle concentration* or *critical micellization concentration,* abbreviated as CMC.

Adsorption amounts of the surfactant are calculated from the slope of the curves in Figure 3.2, and are plotted against concentration in Figure 3.3. The adsorption amount increases gradually with increasing concentration of surfactant and reaches a saturated value at a certain concentration. The concentration at which the adsorption is saturated is much lower than CMC. As already pointed out, the Gibbs' equation cannot be applied to the concentration region above CMC, and the curves are stopped at CMC. The adsorption itself, however, maintains at the saturation value even above CMC, as mentioned previously.

b. Surface Tension of Some Aqueous Solutions of Surfactants

Figure 3.4 shows surface tension-concentration curves of sodium alkylsulfates. Surface tension lowering is limited at about 35 mN/m irrespective of the alkyl chain length. Most water-soluble surfactants show a similar result to the above. The surface tension of liquid *n*-alkanes is much lower, e.g., 21.8 mN/m (at 20°C)

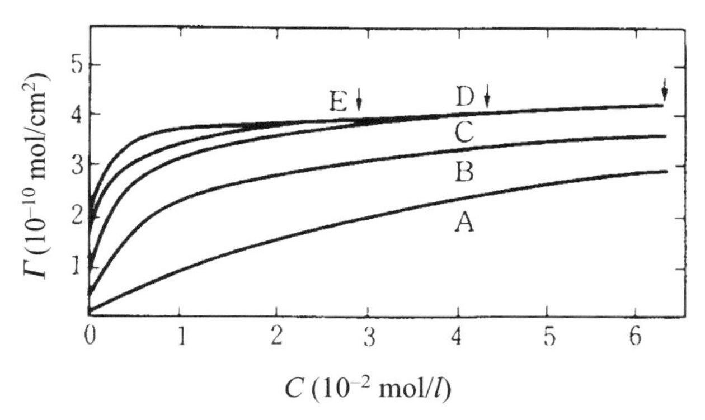

Figure 3.3 Adsorption amount versus concentration curves calculated from the data shown in Figure 3.2 (Ozeki *et al.*, 1978).

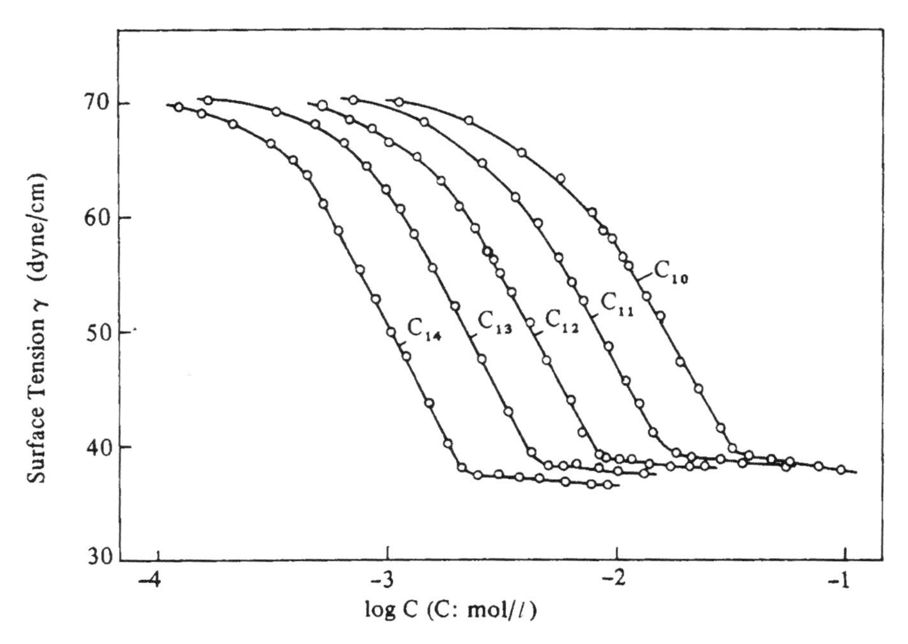

Figure 3.4 Surface tension of an aqueous solution of sodium alkyl sulfates as a function of their concentration. Reprinted by permission of Kao Corp.

for *n*-octane. The surface tension of aqueous solutions of surfactants must be lowered to about 20 mN/m if the hydrophobic alkyl chains are packed with each other at the surface of the solution as closely as those of the liquid alkane molecules. The surfactant molecules may be packed more loosely at the surface of the solution owing to the repulsive interaction between ionic head groups of the molecules. This may be why the surface tension of the aqueous surfactant solutions decreases only up to 35 mN/m. Addition of salt reduces the repulsion between the ionic heads and may make the hydrophobic chains of the surfactant pack more compactly. One can understand from Figure 3.2 that the lowest surface tension obtained at a concentration range higher than CMC decreases with the addition of NaCl.

The surface tension of an aqueous surfactant solution is lower than that of pure water because the adsorption layer of hydrocarbon chains of surfactant molecules has less cohesive energy. Of course, even lower surface tension will be obtained if surfactants having hydrophobic groups with even less cohesive energy are used. For example, fluorocarbon is well known as a compound that has low cohesive energy. Figure 3.5 shows the surface tension-concentration curves of aqueous solutions of some surfactants having a fluorocarbon chain as a hy-

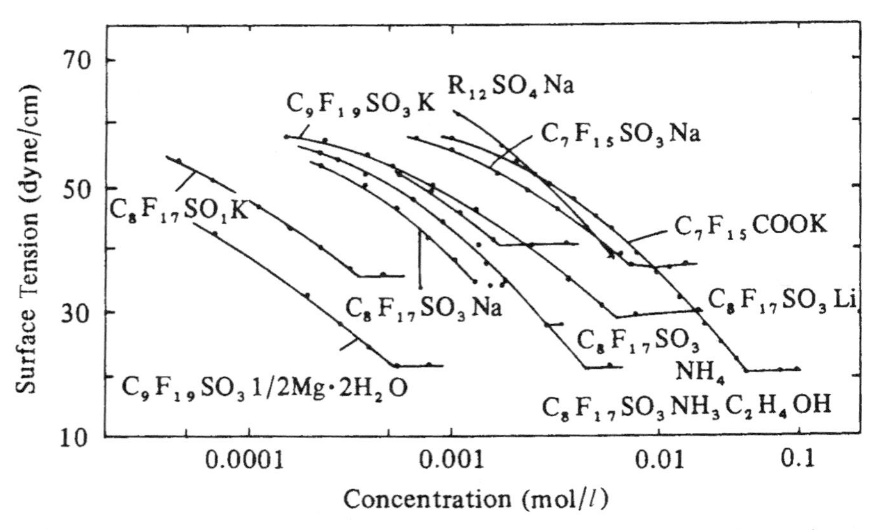

Figure 3.5 Surface tension-concentration curves of some fluorinated surfactants in water. Data of sodium dodecyl sulfate ($R_{12}SO_4Na$) are also included as a reference (Shinoda *et al.*, 1972). Reprinted with permission from *J. Phys. Chem.*, **76,** 909 (1972), American Chemical Society.

drophobic group (Shinoda *et al.*, 1972). We can see that much lower surface tension values are obtained in the solutions of fluorocarbon surfactants than in those of hydrocarbon surfactants.

c. Enhancement of Wetting by Adsorption of Surfactants

Enhancement of wetting is a remarkable phenomenon resulting from the surface tension lowering of water with the dissolution of surfactants. For instance, a water droplet put on a surface of a frying pan coated with Teflon (polytetrafluoroethylene) is ball-shaped, but it changes its shape dramatically—it flattens out—when a small amount of surfactant is dissolved in the water droplet.

The shape (contact angle) of a liquid droplet on a solid surface is determined by the balance of three surface and interfacial tensions, as illustrated schematically in Figure 3.6(a). The horizontal components of three surface and interfacial tensions must be balanced in equilibrium. Then, the following Young's equation is held.

$$\gamma_S = \gamma_{SL} + \gamma_L \cos\theta \tag{3.2}$$

or

$$\cos\theta = \frac{(\gamma_S - \gamma_{SL})}{\gamma_L} \tag{3.3}$$

where γ_S, γ_L, and γ_{SL} are the surface tensions of solid, liquid, and solid/liquid interfacial tension, respectively. θ is the contact angle of the liquid droplet on the solid surface. How does the contact angle change when a surfactant is added to the liquid (say, water)? The surface tension of the liquid (γ_L) and the interfacial tension (γ_{SL}) are depressed by adsorption of a surfactant at the surface of the liquid and solid/liquid interface. Consequently, the right-hand side of Eq. (3.2) becomes smaller and the liquid droplet spreads (the contact angle becomes smaller), being pulled by the surface tension of the solid (γ_S). This is the principle of wettability enhancement by the surface active agents (illustrated in Figure 3.6(b)).

Let us take one more example of this wettability enhancement property: immersion wetting such as the capillarity phenomenon. Figure 3.7 shows the capillary action phenomenon of a liquid. In this case, the liquid penetrates into a capillary of radius r being driven by the imbalance of the surface tension of the solid and the interfacial tension between the solid and the liquid. The force that pulls the liquid into a capillary is written as $2\pi r(\gamma_S - \gamma_{SL})$ since the surface (interfacial) tension is defined as force/unit length (see Eq. (1.1)). We can obtain the

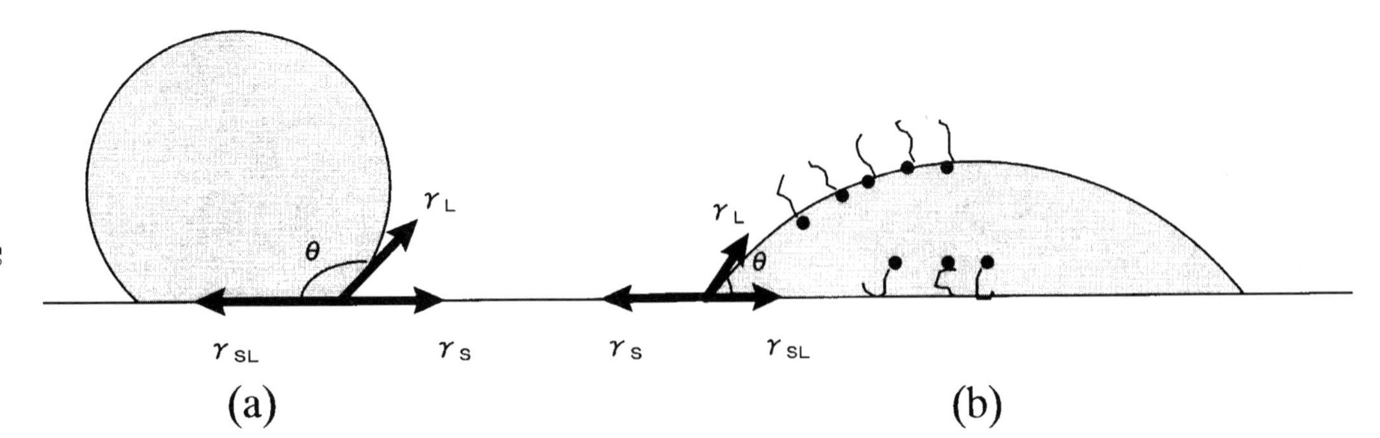

Figure 3.6 Enhancement of wetting by a surfactant. A contact angle of a liquid droplet on a solid surface is determined by the balance of three surface and interfacial tensions (a) and becomes smaller when a surfactant is dissolved in the liquid (b). This is because the surface tension of the liquid and the interfacial tension of the liquid/solid interface are both reduced by surfactant adsorption.

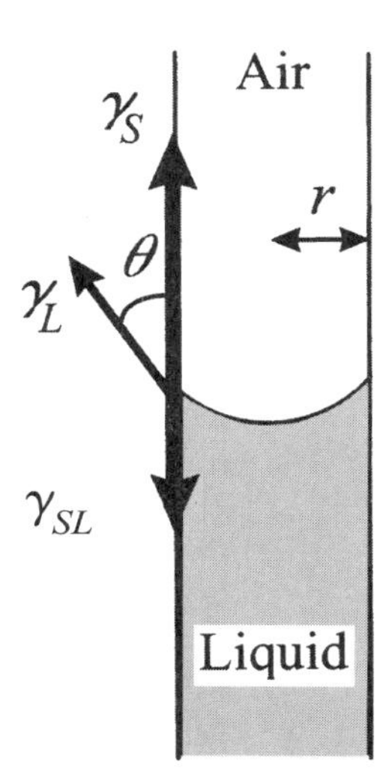

Figure 3.7 A capillary action phenomenon (an immersion wetting). The liquid penetrates into a capillary when θ is less than 90° but comes out when it is greater than 90°.

capillary pressure (ΔP) by dividing the previous force by a cross-sectional area of the capillary. Then,

$$\Delta P = \frac{2\pi r(\gamma_S - \gamma_{SL})}{\pi r^2} = \frac{2(\gamma_S - \gamma_{SL})}{r}. \tag{3.4}$$

Eq. (3.4) indicates that the pressure in the liquid side is lower than that in the air side by the value of ΔP, when the meniscus is concave to the air. A curved liquid surface tends to become flat in order to minimize the surface area. This action of the surface tension of the liquid results in the lower pressure in the liquid side.

Combining Eq. (3.2) with Eq. (3.4), we obtain

$$\Delta P = \frac{2\gamma_L \cos\theta}{r}. \tag{3.5}$$

One can understand that $\Delta P > 0$ when $\theta < 90°$ and $\Delta P < 0$ when $\theta > 90°$. The liquid penetrates into the capillary when $\Delta P > 0$, and the liquid is pushed out of the capillary when $\theta > 90°$. Thus, water does not penetrate into a Teflon and/or polyethylene capillary, but does into a glass one.

The surfactant enhances the wettability of water toward the hydrophobic solid surface with low surface tension. As mentioned previously, the contact angle of water becomes smaller, and then ΔP turns to positive when the surfactant is added to the water. Consequently, water (liquid in general) can penetrate into hydrophobic capillaries. This enhancement of wettability by the surfactant has a variety of applications. For example, hydrophobic powders (say, carbon black) cannot be dispersed into water because the water medium cannot penetrate into their interparticle spaces. Neither can water penetrate into the interfabric spaces of oily dirt on clothing. In such cases, the surfactant contributes dramatically to the dispersing of the carbon black and to the cleansing of oily clothes.

d. Wettability of Solid Surfaces Having a Fractal Structure

The contact angle of a liquid on a flat solid surface (Young's equation (3.3)) is determined by the chemical factor of materials of the liquid and the solid, i.e., the combination between the liquid and solid surfaces. However, there is one more important factor that determines the contact angle: the geometrical (microscopic roughness) factor of the solid surfaces. When the solid surface is rough (Adamson, 1990a), Young's equation can be rewritten as

$$\cos \theta_r = r \frac{\gamma_S - \gamma_{SL}}{\gamma_L} = r \cos \theta, \tag{3.6}$$

where θ_r and r are the contact angle on the rough surface and the roughness factor, respectively. The roughness factor, r, is the surface area magnification factor giving the ratio of actual to projected area. Eq. (3.6) means that both forces pulling to the solid surface and to the solid/water interface in Figure 3.6(a) become larger by r times, since the forces are expressed by the surface tension × surface area. One can understand from Eq. (3.6) that the wettability is enhanced by the roughness of the surface; i.e., the wettable surface becomes more wettable and the repellent surface becomes more repellent when surface is rough since r is always greater than unity.

A fractal surface is an ideal surface that gives the largest surface area magnification factor (Mandelbrot, 1982) and may provide a super-wettable or a super-repellent surface for some liquids. When the wettability of fractal surfaces was studied theoretically and experimentally, a super-water-repellent surface indeed was realized (Onda *et al.*, 1996; Shibuichi *et al.*, 1996). The roughness factor, r, in Eq. (3.6) can be expressed as $(L/l)^{D-2}$ when the surface roughness of the solid is fractal. Then Eq. (3.6) is rewritten as

$$\cos \theta_f = \left(\frac{L}{l}\right)^{D-2} \cos \theta, \tag{3.7}$$

where θ_f is the contact angle on the fractal surface, L and l are the largest and the smallest size limits between which the fractal (self-similar) structure holds, and D is the fractal dimension. Enhancement of wettability becomes greater when the self-similarity holds in a wider range (L is larger and l is smaller) and the fractal dimension is larger.

Equation (3.7) is derived assuming complete contact between the liquid and solid surfaces. However, the liquid (water for example) cannot penetrate into tiny cavities on the hydrophobic fractal solid surface. On the other hand, on hydrophilic surfaces water vapor adsorbs and condenses into liquid in small hollows before its saturated vapor pressure (capillary condensation). Taking these factors into account, the wettability of a fractal surface is given as shown in Figure 3.8. This theoretical result predicts that super-water-repellent or super-wettable surfaces are possible if the surfaces are fractal.

The super-water-repellent fractal surfaces actually have been realized with a wax (alkylketene dimer). Figure 3.9 shows a water droplet having a diameter of ~1 mm on the fractal surface of the wax. One can see the almost perfect water-repellency (contact angle = 174°) of the solid surface without any fluorination treatments (Onda *et al.*, 1996; Shibuichi *et al.*, 1996).

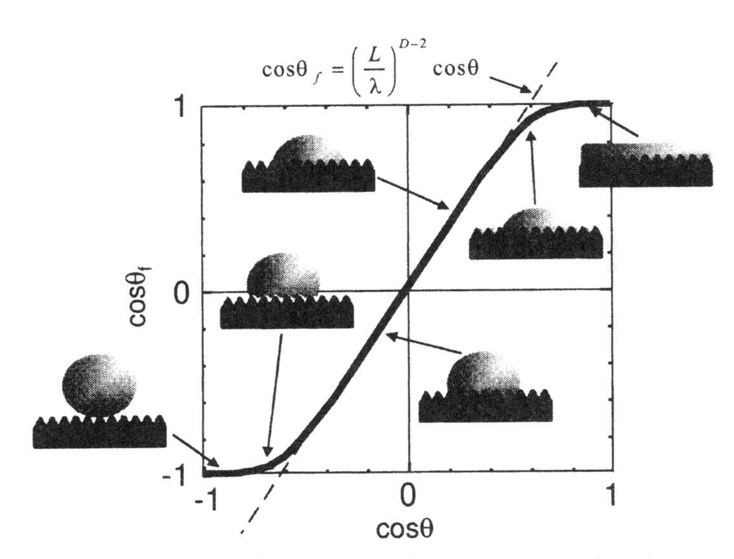

Figure 3.8 Theoretical curve of the wetting enhancement by the fractal structure of a solid surface. θ and θ_f are the contact angles of a liquid on a flat and a fractal surface of the solid, respectively. The straight line passing through the original point indicates Eq. (3.7).

Figure 3.9 A water droplet with a diameter of 1–2 mm placed on a super-water-repellent surface made of alkylketene dimer (an organic wax). The contact angle is 174°.

When we combine the effects of the fractal structure of surfaces and the very low surface energy (~6 mN/m) of the trifluoromethyl group, even super-oil-repellent surfaces are possible (Tsujii *et al.*, 1997b). The authors have succeeded in making a surface on which some oils (such as rapeseed oil) roll around without attaching to the surface. The contact angles of some polar oils on the surface were greater than 150°. The super-oil-repellent surface was prepared by treating an anode-oxidized aluminum plate having the fractal dimension of 2.16 with a fluorinated monoalkylphosphate. But this surface is still far from ideal since its critical surface tension is 12–13 mN/m, which is just an intermediate between Teflon (18.5 mN/m) and the trifluoromethyl group (~6 mN/m).

3.2.2 FOAMING AND DEFOAMING

a. Foam Stabilization Mechanism

Foaming and defoaming are two of the typical phenomena which are caused by the adsorption of a surfactant at an air/water interface. People often believe that surfactants stabilize the foams due to their surface tension lowering action on water. The logic is like this: "Surface area increases very much in the foaming process." → "Surface free energy increases because of the increase of surface area." → "The increment of surface free energy by foaming can be reduced by lowering the surface tension of water by the addition of a surfactant." → "Foaming can be easily stabilized by the addition of a surfactant." But this logic is wrong. It fails to take into account the fact that the energy needed to make the foams is provided externally. For example, a washing machine provides such energy in the process of washing clothes, human hands provide it in the process of shampooing, and so on. Furthermore, it is easily understood from a simple calculation that the required free energy to make the foams is quite small compared with that provided by the external forces. If low surface tension in liquids led to high foaminess, most of the organic liquids (such as hydrocarbons, ethers, etc.) would all have a high foaming property. But this is not the case at all.

So if surface tension lowering is not the reason why the surfactant stabilizes the aqueous foams, what then is the mechanism of foam stabilization by surfactants? Figure 3.10, a schematic illustration of foam film, will help provide the answer. Three bubbles are in contact each other at a point p, which is called the plateau's border. An expanded plateau's border is drawn in Figure 3.10(b). At the plateau's border the surface of the foam film is in a concave shape, and the capillary pressure is lower than that in the film portion having parallel surfaces. Then the liquid inside the foam film flows into the plateau's border. If there are many plateau's borders, the liquid in the upper borders flows toward the bottom ones

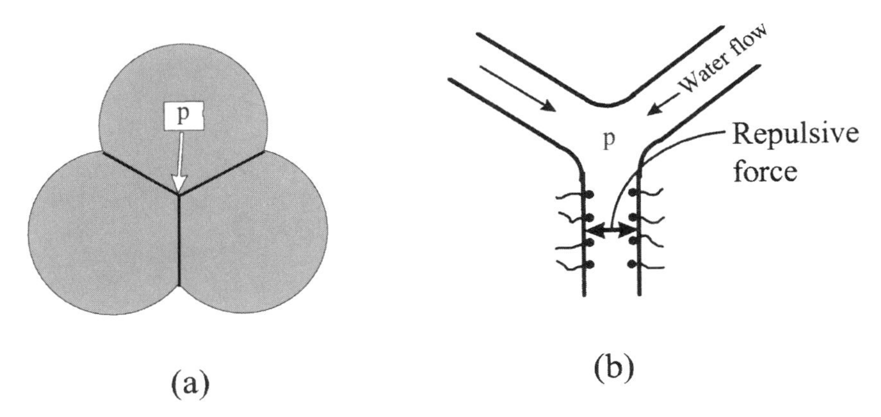

Figure 3.10 A schematic illustration of foam films. A plateau's border (p) is formed at the contact point of three bubbles (a) and its expansion (b).

due to gravity. This phenomenon is called *drainage of foams* and makes the foam films thinner. Thus, the foam films become thinner and thinner and finally break down if forces opposing this thinning pressure do not interfere. Pure liquids do not have any such opposing forces and do not provide stable foams, even if they possess low surface tension.

Surfactant molecules adsorb at the surface of the foam films in aqueous surfactant solutions. In ionic surfactant solutions, two electric double layers of both surfaces of a foam film give an electric repulsive force between the surfaces and oppose the thinning pressure (see Figure 3.10(b)). This is why ionic surfactants can stabilize the aqueous foams. Nonionic surfactants and polymers like proteins are also able to stabilize foams in terms of steric repulsive forces. These repulsive forces are also important in the stability of emulsions and suspensions, and will be discussed in detail in Section 3.2.4.

b. Mechanism of Defoaming and Defoaming Agents

To extinguish the foams, the adsorbed molecules that give the opposing force to the thinning pressure must be removed from the surface of the foam film. Defoaming (or antifoaming) agents show such action; i.e., they sweep the molecules making the foams stable out of the surface of the liquid films. Thus, the necessary prerequisites for defoaming agents are (1) they must be more surface active and adsorb and/or spread on the surface of foam films more strongly than the foam-stabilizing compounds and (2) they must not have any functional groups that give the opposing forces to make the foam stable. Higher alcohols, fatty acid esters, phosphate esters, poly(propylene glycol), and silicone oil emulsions are typical defoaming agents; these are not very water-soluble but do spread easily on water surfaces. Obstinately stable foams once formed and drained fully can often be destroyed by water-soluble solvents such as ethanol. In this case, the foam films themselves, consisting of the foam-stabilizing compounds and water, are dissolved into the solvent.

c. Stratification in the Thickness of Foam Films

One will find an interesting phenomenon if one observes carefully a hemispheric soap bubble on a glass plate or a soap film spread in a frame. At an early stage, the soap film shows an iridescent color due to the interference of visible light by the submicrometer film thickness. In due time, a dark spot suddenly appears on the top of the hemisphere or of the frame and then expands to the bottom. In the next step, a completely transparent spot appears and again expands. These dark and/or transparent portions of the soap film have a thickness less than the wavelength of visible light and are called *black film.*

The most interesting point of the previous observation is the discontinuous stepwise thinning of the soap film. This stepwise thinning has been observed from the beginning of this century, but its mechanism only has been made clear recently (Nikolov *et al.*, 1989; Kralchevsky *et al.*, 1996). Discontinuous thinning takes place only when monodispersed particles—such as surfactant micelles and monodispersed latices—are present in the solutions. Monodispersed particles form a somewhat regular layered (stratified) structure, as shown in Figure 3.11 (Nikolov *et al.*, 1989). These layers of monodispersed colloids come out of the soap film one layer by one layer, and then the thinning occurs discontinuously. In addition, a free energy barrier is present when an empty spot is nucleated in an outermost layer of homogeneous distribution of particles (Kralchevsky *et al.*, 1990). Thus, the stratification phenomena of monodispersed particles in the liquid films can contribute to the stabilization of the foam films.

3.2.3 EMULSIONS

Emulsification is the most important phenomenon and/or technology caused by the adsorption of surfactant at liquid/liquid interfaces. There are many good books on emulsions (Becher, 1957; Sherman, 1968; Becher, 1983, 1985, 1988; Larsson and Friberg, 1990; Sjöblom, 1996), and so we focus here on the basic principles of emulsions from the viewpoint of surfactant adsorption.

a. Type of Emulsions

Emulsion is defined as a macroscopically homogeneous mixture of two (or more) kinds of immiscible liquid. It is, of course, not a real solution. Typical immiscible liquids are water and oil, and for these there are two types of emulsion. One is oil in water (O/W), in which oil droplets are dispersed in a continuous phase of water. The other is water in oil (W/O), which has a continuous oil phase containing water droplets. Milk and mayonnaise are examples of O/W emulsions; butter and margarine are W/O ones.

Emulsion is not a thermodynamically equilibrium state, and the type of emulsion depends on the procedures of emulsion making. When we make the emulsions by different methods, the emulsion types can be different from each other even if they have exactly the same composition and are kept at the same temperature. The governing factors are, for example, which phase the surfactant is dissolved in (water phase or oil phase), whether the water phase is mixed into the oil

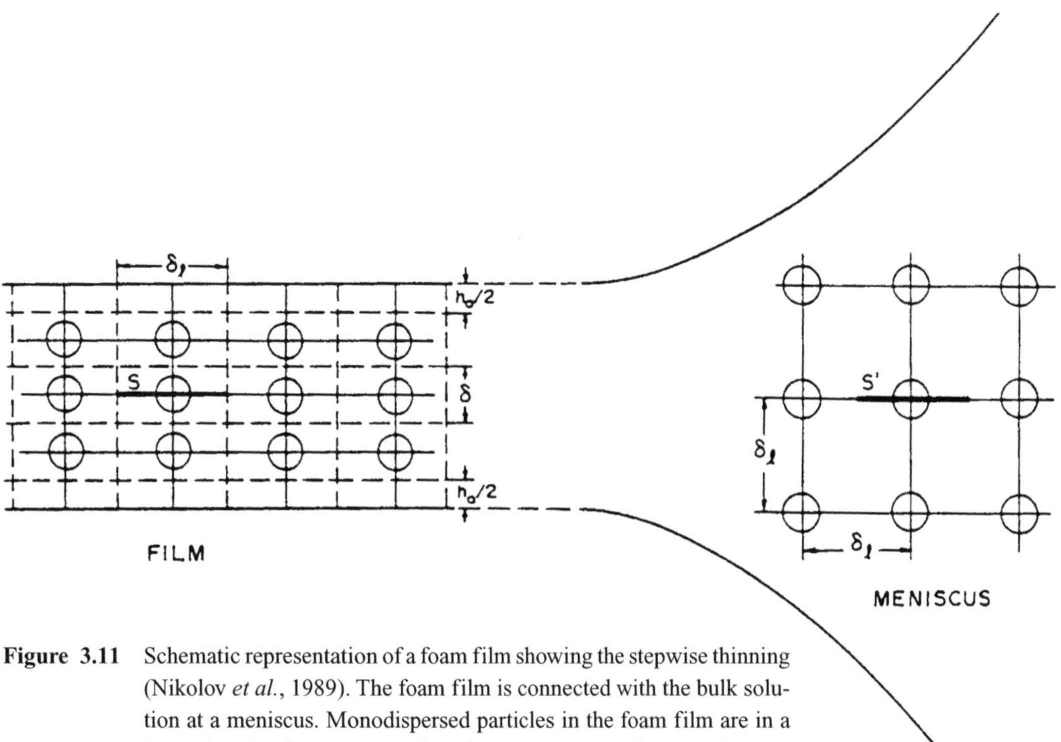

Figure 3.11 Schematic representation of a foam film showing the stepwise thinning (Nikolov *et al.*, 1989). The foam film is connected with the bulk solution at a meniscus. Monodispersed particles in the foam film are in a layered, ordered structure, and the layers are squeezed out one by one. This thinning mechanism results in the stepwise change in the thickness of the foam film. The average distance between the particles in the film (δ) is shorter than that in the bulk solution (δ_l) since the spontaneous thinning pressure is acting between two surfaces of the film.

phase or vice versa, the history of temperature change during emulsification, and so on. There are several empirical rules determining the type of emulsion.

1. *Emulsifier:* The phase into which the emulsifier dissolves more has a tendency to become the continuous phase. In other words, O/W emulsion is readily made when a hydrophilic surfactant (emulsifier) is used, and W/O emulsion is easily obtained by a hydrophobic one. This rule is known as *Bancroft's rule* (Bancroft, 1913, 1915).
2. *Volume ratio of two liquids:* One can easily understand that the liquid of larger volume tends to become the continuous phase.
3. *Temperature:* The temperature effect is particularly large in emulsions stabilized with nonionic surfactants (emulsifiers). In this case, W/O emulsions are readily formed at higher temperatures.
4. *Vessel material:* Water has a tendency to become the continuous phase when a hydrophilic vessel such as a glass is used, and vice versa. It is reasonable that the liquid more wettable to the vessel wall more easily becomes the continuous phase.

These rules are empirical and qualitative ones; only a few have been theoretically verified. Overall, emulsification is still a technology of some art rather than of science.

b. Emulsion Stability

Emulsion is not stable in a thermodynamic sense because of its high interfacial energy (area) between the two phases of water and oil. The emulsions, therefore, must be separated into two bulk phases sooner or later to minimize the interfacial area. So the stability of emulsions is defined kinetically, which just means that the rate of separation is slow in stable emulsions. The factors contributing to the kinetic stability of emulsion will be discussed later; we mention here the practical significance of emulsion stability.

Separation of emulsions proceeds by two steps. In the first step, the emulsion droplets are floating up (or settling down) owing to the gravity. In the case of O/W emulsions, oil droplets are floating up and the water phase is separated out in the bottom of the vessels. We call this step *creaming*. In the creaming step, emulsion droplets are not changed in size. The second step is the coalescence of emulsion droplets. In this step, the size of emulsion droplets becomes larger and larger, until finally bulk separation takes place. Figure 3.12 illustrates these steps. In the case of W/O emulsions, the opposite phenomenon of course occurs.

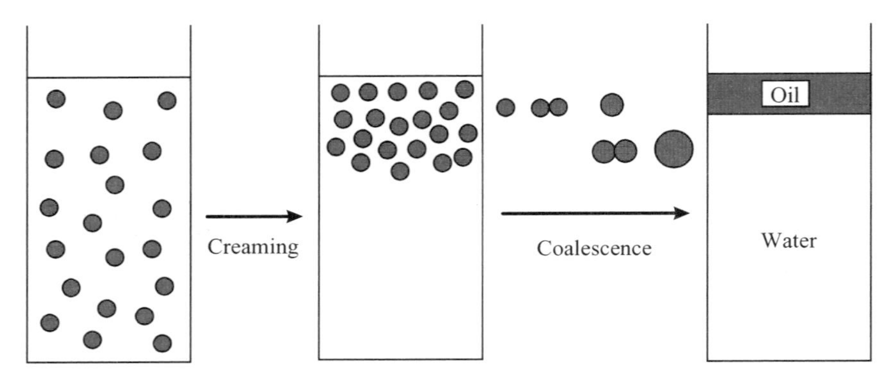

Figure 3.12 Separation process of emulsions (creaming and coalescence).

We have only simple methods for avoiding the creaming phenomenon. One is to make the emulsion particles small enough to be dispersed by Brownian motion; the particle size necessary for this purpose is less than 1 μm. The other method is to make the dispersing medium viscous enough to prevent the particles from floating up or settling down. High polymers are often utilized to obtain this viscous medium. A viscoelastic solution having a yield value in a stress-strain curve is particularly useful to prevent the creaming process. Liquid crystal emulsification is a special method that provides a strong viscoelastic continuous medium of the liquid crystal and gives an extremely stable emulsion. This unique method will be described in Section 4.4.2.*a*. To avoid the second step, i.e., the coalescent process of emulsion droplets, surfactants or emulsifiers play very important roles. Let us discuss this problem in the next section.

c. HLB of Surfactants and Emulsion Stability

To obtain the stable emulsions against coalescence, surfactant molecules must adsorb efficiently at the interfaces between the water and oil phases. The surfactant having too hydrophilic a nature dissolves into the water phase, and one having too hydrophobic a nature dissolves into the oil phase. In either case, the surfactant molecules may not adsorb efficiently at the interfaces. Consequently, the hydrophilic and hydrophobic nature must be balanced in the surfactant molecule for it to work well as a good emulsifier. This hydrophilic and hydrophobic (lipophilic) balance of a surfactant is called *HLB* (hydrophile/lipophile balance). HLB is given as a number to each surfactant and used as a guiding criterion to select suitable emulsifiers.

These numbers can easily be found in a textbook of emulsions, and some of them are summarized in Table 3.1. The HLB number was defined by Griffin for the nonionic surfactant of poly(oxyethylene) type (Griffin, 1954) as

$$\text{HLB number} = \frac{E}{5}, \tag{3.8}$$

where E is the weight % of the poly(oxyethylene) chain in the nonionic surfactant molecule. So the largest possible HLB number is 20. The HLB number of a mixed surfactant is given by the weight-averaged value.

Table 3.1 HLB Values of Typical Surfactants and Required HLB of Some Typical Oils

Surfactant	***HLB***
Sodium dodecyl sulfate	40
Sodium oleate	18
Polyoxyethylene (p = 6)* dodecyl ether	11
Polyoxyethylene (p = 20) dodecyl ether	16
Sorbitan mono-oleate	4.3
Sorbitan mono-stearate	4.7
Sorbitan tri-stearate	2.1
Polyoxyethylene (p = 5) sorbitan mono-oleate	10
Polyoxyethylene (p = 10) sorbitan mono-oleate	13
Polyoxyethylene (p = 20) sorbitan mono-oleate	15
Sugar mono-laurate	13
Stearoyl mono-glyceride	3.8
Propylene glycol mono-laurate	4.5

*Polyoxyethylene (p = m): m oxyethylene units adduct

Oil	***Required HLB O/W Emulsion***	***W/O Emulsion***
Hexadecyl alcohol	15	
Methyl salicylate	14	
Palm oil	13	
Kerosene	12	6 ~ 9
Cottonseed oil	12	
Silicone oil	11	
Oleic acid	11	7 ~ 11
Paraffin oil	11	6 ~ 9
Hydrogenated tallow	11	
Isopropyl palmitate	10	
Lanolin dehydrated	10	8
Polyethylene wax	9	

There are many kinds of oils, from nonpolar to polar ones. Furthermore, more hydrophobic oils—such as silicone (polydimethylsiloxane)—are now frequently used in toiletries and cosmetics. To emulsify oils of such a variety of hydrophobicity, we have to select the surfactant having the most suitable HLB number for the target oil. The HLB number of the best surfactant to emulsify a certain oil is called the *required HLB* of the oil.

We can estimate the unknown HLB number of some surfactant experimentally if we have a surfactant with a known HLB number and an oil with a known required HLB. To do this, we emulsify the oil by changing the mixing ratio of two kinds of surfactant and observe the stability of the emulsions. We determine the mixing ratio of surfactant that gives the most stable emulsion, and the HLB number of this mixture must be equal to the required HLB of the oil. Then we can extend the HLB numbers defined at first for nonionic surfactants to ionic and any other types of surfactants.

The concept of HLB was proposed and developed for surfactants and oils of hydrocarbon type. No one yet knows whether this HLB concept can be applied to new types of ingredients such as silicone oils and surfactants or fluorinated surfactants and oils. Because such ingredients are now extensively used in the toiletry and cosmetic industries, research on the application of the HLB concept is necessary.

So far we have discussed how to make the surfactant molecules adsorb at the oil/water interfaces efficiently. We now discuss how the adsorbed surfactant molecules stabilize the emulsions against coalescence. This problem is essentially the same as that of the stabilization mechanism of suspensions and foams, and will be described in detail later (Section 3.2.4).

d. Phase Inversion Temperature and the Type of Emulsion

The HLB number is specific to each surfactant, as can be easily understood from its definition. Emulsion stability, however, changes considerably with changing conditions such as temperature, pH, electrolyte concentration, etc., even if we use the same surfactant and oil. So we have to recognize that the HLB number is more or less variable depending on the conditions. Thus, the phase inversion (or HLB) temperature method is the way to determine the hydrophilic/hydrophobic balance of the surfactant from the solution properties of emulsions, not from the chemical structure of the surfactant.

Figure 3.13 shows a phase diagram of the water/cyclohexane system containing 7 wt % of polyoxyethylene ($\overline{p} = 9.7$) nonylphenyl ether (Shinoda and Becher, 1978b). The diagram looks complicated, but the essential curve is that of the phase

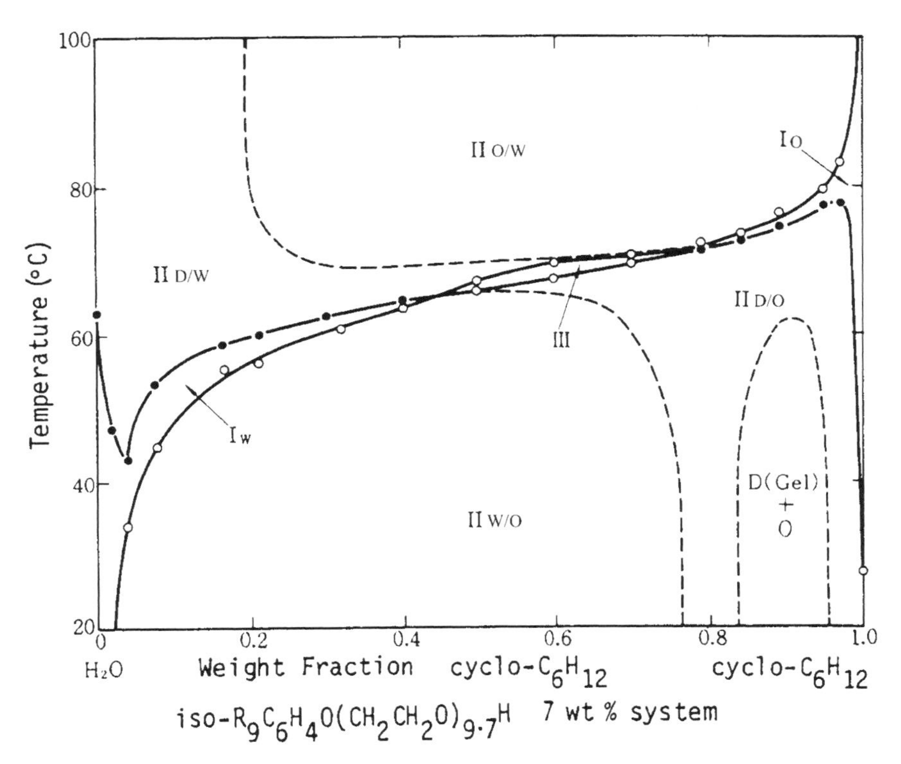

Figure 3.13 Phase diagram of the system of water/cyclohexane containing 7 wt % polyoxyethylene (p = 9.7) nonylphenyl ether (Shinoda and Becher, 1978b). Reprinted from "Principles of Solution and Solubility", p. 192 courtesy of Marcel Dekker, Inc.

inversion temperatures represented by filled circles (•). O/W (W/O) emulsion is preferably formed below (above) this curve of the phase inversion temperatures. For example, the polyoxyethylene-type nanionic surfactant behaves as a water-soluble agent at low temperatures and as an oil-soluble one at higher temperatures, as will be mentioned in Section 3.3.2. In other words, the HLB of this surfactant becomes smaller with increasing temperature. As a result, this surfactant is suitable for O/W emulsions at low temperatures and for W/O emulsions at higher ones. At a certain intermediate temperature, called the *phase inversion temperature* (PIT), the emulsion type transforms from O/W to W/O.

Phase inversion occurs with temperature change in the case of nonionic surfactants. One can observe the phase inversion also by changing the mixing ratio of hydrophilic and hydrophobic surfactant as well as by adding salt to ionic emulsifier systems.

The phase inversion temperature certainly defines the hydrophile/lipophile balance exactly, but cannot assign one specific value to one surfactant. Therefore, from a practical point of view, the HLB number system is more convenient than the PIT system for the selection of a suitable surfactant to emulsify some target oil. The most practical way to select the best emulsifier system is to first choose a surfactant having the same HLB number as the required HLB of the target oil and then adjust the surfactant system properly by watching the stability of obtained emulsions.

3.2.4 DISPERSIONS AND FLOCCULATIONS/COAGULATIONS

Wetting and dispersion are the two main phenomena resulting from the adsorption of a surfactant at solid/liquid interfaces. The wetting phenomenon was discussed in Section 3.2.1 since wetting is related to adsorption at liquid/gas interfaces. Accordingly in this section, we will focus on the dispersion phenomenon and will discuss wetting only as it relates to the dispersion phenomena.

a. Requisite of Dispersion (Immersion Wetting)

Hydrophobic powders such as soot cannot be dispersed in water. They float up even if pushed into a water medium because water does not penetrate into the interparticle spaces of the powders. As described in Section 3.2.1.*c*, penetration of water is governed by immersion wetting. Immersion wetting is determined by the balance between the surface tension of the solid and the solid/liquid interfacial tension. In the case of dispersion of soot into water, the interfacial tension is larger than the surface tension of the solid; thus the water cannot penetrate into the small spaces between the powders to divide them into individual particles. A surfactant can change such situations dramatically. Surfactant molecules adsorb at the solid/water interfaces and make the solid surface hydrophilic. As a result, the interfacial tension between the solid and the water becomes smaller than the surface tension of the solid. The water can now penetrate into the interparticle spaces and disperse the powders.

b. Stabilization of Dispersion by Electrostatic Repulsion (The DLVO Theory)

Dispersion is defined as a homogeneously mixed state of small particles in a continuous liquid medium. A large interfacial area between the particles and the liquid medium leads the system to a thermodynamically unstable state. So the stability of dispersion is defined kinetically, just like that of the emulsion systems. A surfactant can contribute by slowing the coagulation rate and making the dispersion stable. Anionic surfactants, for example, adsorb on the particle surface and give the particles an anionic charges. Electrostatic repulsive interaction due to the given charges prevents the particles from coagulation.

Stabilization of dispersions by electrostatic repulsion was theoretically established by Verwey and Overbeek (1948) and is widely known as the DLVO theory. (DLVO are the initials of the authors who named this theory—Derjaguin, Landau, Verwey, and Overbeek). One can find the detailed treatments of this theory in the literature (e.g., Hiemenz and Rajagopalan, 1997; Adamson, 1990b), so only the essential points are mentioned in this section. According to the DLVO theory, the interaction potential energy, V, between two colloidal particles is the sum of the electrostatic repulsive, V_R, and the van der Waals attractive, V_A, interactions;

$$V = V_R + V_A. \tag{3.9}$$

The mathematical expression for these interaction potentials depends on the shape of the particle. In the case of platelike particles, the potentials are expressed as

$$V_R = \frac{64nkT}{\kappa} \gamma^2 \exp(-2\kappa d) \tag{3.10}$$

and

$$V_A = -\frac{A}{48\pi d^2}, \tag{3.11}$$

where γ and κ are written as

$$\gamma = \frac{\exp(\omega/2) - 1}{\exp(\omega/2) + 1} \qquad \omega = \frac{ze\Psi_0}{kT} \tag{3.12}$$

$$\kappa = (8\pi ne^2z^2/\epsilon kT)^{1/2}, \tag{3.13}$$

where n and z are the counterion concentration and its valence number, respectively, d is the distance between two surfaces of the particles, e is the electronic charge, ϵ is the dielectric constant of the medium, k is the Boltzmann constant, T is the absolute temperature, A is the Hamaker constant, and Ψ_0 is the electrostatic potential at the surface of the particle. The interaction potentials above are changed to the following equations when the particles are spherical (with radius a).

$$V_R = \frac{\epsilon a \Psi_0^2}{2} \ln\{1 + \exp(-\kappa H_0)\} \quad (3.14)$$

$$V_A = -\frac{aA}{12H_0}, \quad (3.15)$$

where H_0 is the distance between the nearest surfaces of the two spherical particles.

It is not easy to imagine the physical image of the interactions from the above mathematical expressions. Thus, we provide a physical model of this theory for platelike particles in Figure 3.14a. Most of the solid surface bears negative charges in aqueous solutions resulting from the adsorption of specific ions in the solution and/or the dissociation of some functional groups at the surface of the particles. The counterions of the same number as that of the surface charge are, of course, present in the solution side. These counterions are attracted by the surface charge but, on the other hand, diffuse away from the surface by thermal movement. As a result, an equilibrium distribution (Boltzmann distribution) of counterions is attained (Figure 3.14b), and the diffuse electric double layer is formed. The electric neutrality condition is broken only in the interfacial phenomenon, and the charge separation between the solid surface and the solution takes place. The distribution of counterions and the accompanying potential change at the vicinity of the interface are determined by solving self-consistently the two simultaneous equations, i.e., the Boltzmann and the Poisson equations. The Boltzmann equation describes the distribution of ions (charges) under some given electric potentials, and the Poisson equation expresses the potential in the presence of some electric charges. Procedures for solving these simultaneous equations can be found in the literature (e.g., Hiemenz and Rajagopalan, 1997; Adamson, 1990b), and the results are given in Figure 3.14b and 3.14c. The effect of surface potential as well as the counterion distribution, which is denser than that in the bulk solution, extend to the distance of 1–100 nm from the solid surface, depending on the value of κ. The quantity $1/\kappa$, called the *Debye length,* is the measure of the thickness of the above counterion distribution.

When two colloidal particles bearing the electric double layer around them come near each other, the concentration of counterion becomes higher between the two particles than in the bulk solution. Solvent molecules come into this domain to dilute the high concentration of counterion, which results in the repulsive force between the particles. In other words, the osmotic pressure of condensed counterions is the origin of electrostatic repulsion between the two

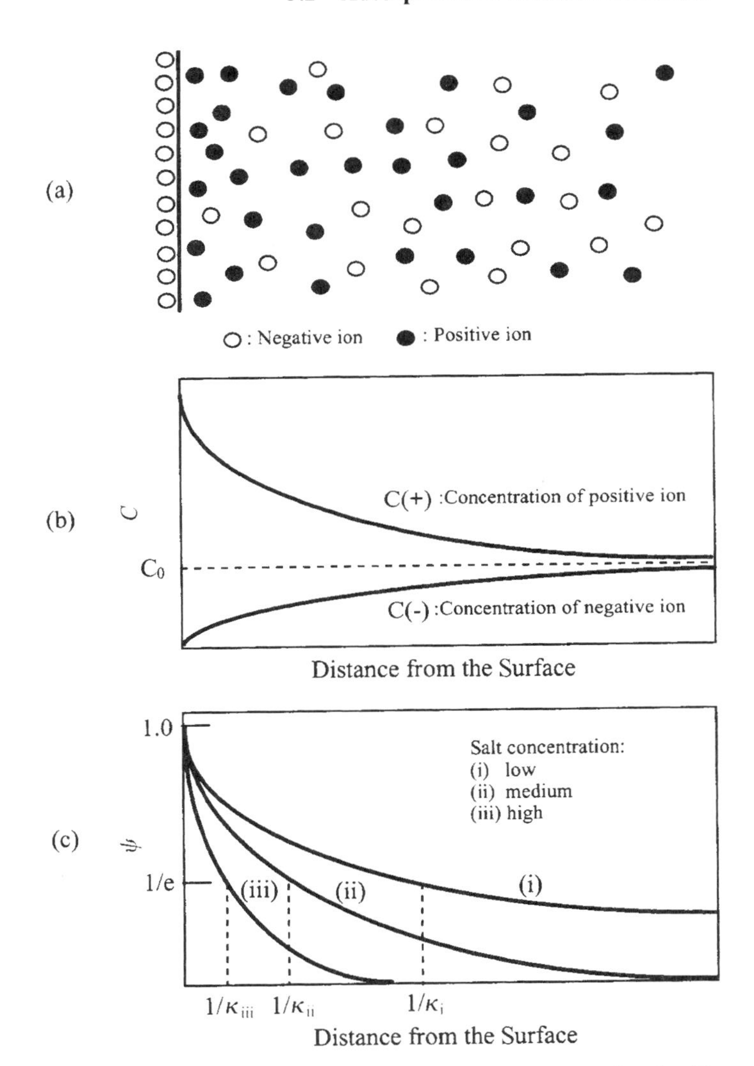

Figure 3.14 Physical image of the diffuse electric double layer of a platelike particle. Schematic illustration of positive and negative charges at the vicinity of the surface (a), charge distribution (b), and electric potential change (c) as a function of distance from the surface.

particles. One may wonder why this is so, but it can be verified by another derivation of the repulsive interaction. In this other derivation, we assume an imaginary state for two particles of zero electric charge and then build up stepwise an infinitesimally small quantity of charge on the particles. In each step, counterions are allowed to distribute in equilibrium. We calculate the electric potential energy in the above procedures until the surface potential reaches the value of Ψ_0. The potential energy is, of course, positive because this procedure is a charge-up process of two condensers. The potential is more positive when two condensers are put closer each other, which results in the repulsive interaction between two particles. These two explanations seem to be completely different from each other, but they lead to exactly the same conclusion.

Let us move to the attraction term. The van der Waals attractive potential is known to be reciprocally proportional to the 6th power of the distance between two molecules. But the attractive interaction between two particles differs in power depending on the distance as expressed in Eqs. (3.11) and (3.15). In the case of particle-particle interactions, the van der Waals potential acting at each point (molecule) between the two particles must be integrated with respect to whole planes in platelike particle systems or whole volumes in spherical particle ones. Consequently, the power of the distance becomes smaller, as shown in the above equations.

The sum of the repulsive and the attractive potentials provides the total interaction potential curve. Figure 3.15 shows some examples of these potential curves. The total potential becomes larger up to a maximum and then falls down in a deep ravine when two particles are approaching each other. The maximum in the potential energy is important for the stability of dispersions. If this potential maximum is high enough compared with thermal energy, the dispersion must be stable. We can regard the dispersions as practically stable when the potential maximum is higher than 15 kT. On the other hand, the colloidal particles easily coagulate when potential barrier is low enough or none. The stability of dispersions is not thermodynamic, as pointed out previously. The most stable state in the deepest potential energy curve is the coagulated one, as easily seen in Figure 3.15. We can see that the stability of dispersions is kinetic, which gives the slow coagulation rate of particles.

Ionic surfactants can contribute to this kinetic stabilization of dispersions since the adsorption of the agents give some electric charges to the colloidal particles. Oligomeric and polymeric dispersants, in particular, are frequently designed to put electric charges to the adsorbents. Their applications in a number of fields will be mentioned in the next chapter.

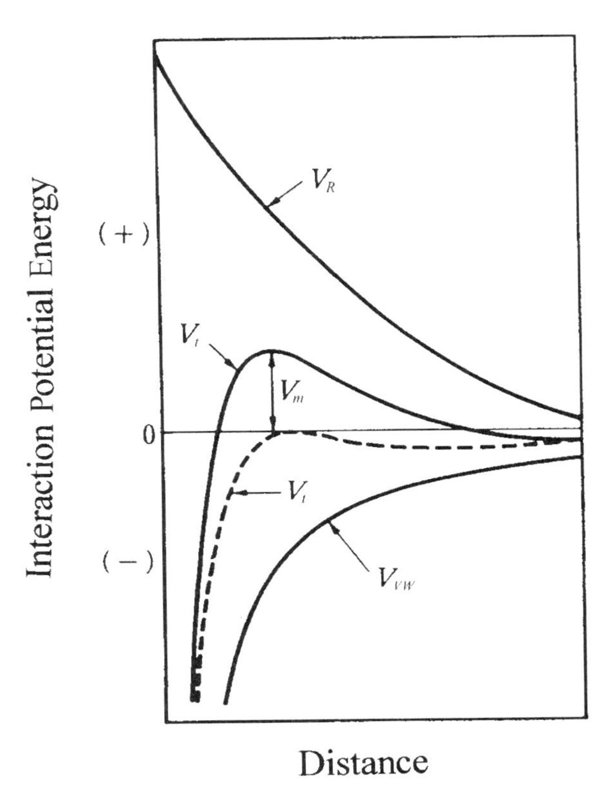

Figure 3.15 Interaction potential energy between two particles as a function of distance. V_R, V_{VW}, and V_t are the electrostatic repulsive, the van der Waals attractive, and the total potentials, respectively. The V_t curve (dotted line) indicates the total potential when the repulsive interaction is small. V_m is a potential barrier to overcome when the two particles come into contact with each other.

c. Stabilization of Dispersion by Adsorbed Polymers (Protective Colloids)

There is one more important method for stabilizing dispersions. This well-known method utilizes the lyophilic colloids, called *protective colloids,* to stabilize other lyophobic colloidal dispersions. The lyophilic colloid is a solute having a diameter of colloidal size, and its dispersing system is a thermodynamically true solution of polymers. Protein solutions and surfactant micelles are some examples of such lyophilic colloids.

Stabilization of dispersions by the protective colloids is now understood as the steric stabilization by adsorbed polymers (Sato and Ruch, 1980; Napper, 1983). Figure 3.16 illustrates schematically the mechanism of steric stabilization. Polymer molecules adsorbed at the surface of the particles take a random coil conformation similar to that in bulk solutions (Figure 3.16a). The most stable conformation is deformed when two adsorbed polymer layers come in contact with each other on approaching the two particles (Figure 3.16b). Further, the polymer concentration becomes higher in the overlapped region than in the bulk phase. Both effects result in the repulsive interaction between the colloidal particles bearing the adsorbed polymer molecules. The former effect gives a repulsion like an entropic rubber elasticity. The latter effect is due to the osmotic pressure of the polymer molecules. These steric repulsion forces are comparable to electrostatic repulsive forces, and they are strong enough to stabilize the col-

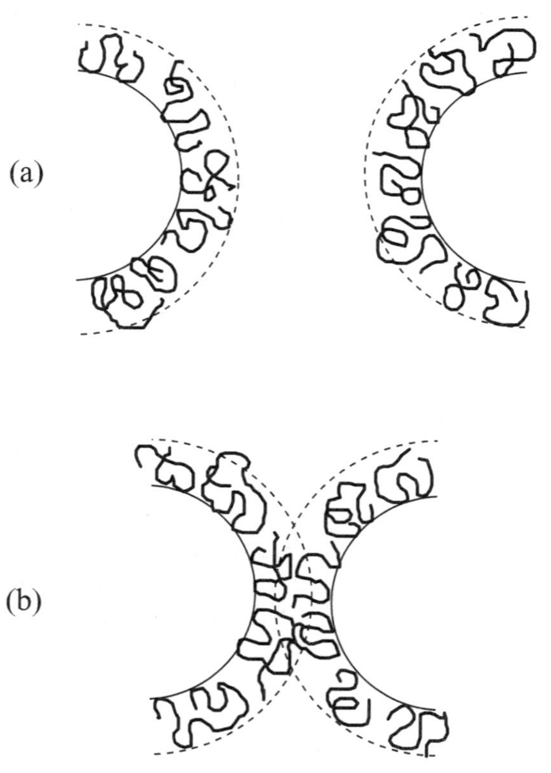

Figure 3.16 Repulsive interaction between two adsorbed layers of polymers (steric stabilization).

loidal dispersions. Polyelectrolytes such as proteins and ionic polysaccharides are often used as protective colloids. In this case, of course, the ionic charges contribute to the stabilization of colloids together with the steric mechanism.

d. Direct Measurement of Surface Forces

The interaction potential energy between two colloidal particles is always expressed as a function of their distance. The potential (force) versus distance curves, however, could not be observed directly until the recent advent of the surface force apparatus (SFA), an epoch in this research field (Israelachvili, 1991; Israelachvili and Adams, 1978). The cleaved mica surface used as the surface for SFA measurements is atomically smooth. The SFA and the mica surfaces under crossed cylinders geometry are shown in Figure 3.17. In this apparatus two mica surfaces can be approached up to the distance of 0.1 nm, and the force between the two surfaces can be directly measured. The distance between the two surfaces is precisely controlled by a piezoelectric crystal and is measured by multiple beam interference fringes of light (Israelachvili, 1973). Force measurements are carried out by a very sensitive spring. A number of interesting and even surprising results have been observed since the SFA method was invented.

Figure 3.18 shows a surface force versus distance curve of two mica surfaces in an aqueous KCl solution of 1.0 mM (Pashley and Israelachvili, 1984). The experimental results agree excellently with the theoretical curve calculated by the DLVO theory in the distance range larger than 2 nm. The results are surprising, however, when two surfaces come closer to each other. The surface force starts to oscillate with respect to distance, and the periodic distance coincides well with the diameter of a water molecule. This interesting result is thought to show that the molecular layers of water between the two mica surfaces are squeezed out one by one. Similar results have been obtained in many organic solvents (Israelachvili, 1991).

Direct measurements of the steric repulsion force—the force acting between two mica surfaces bearing the adsorbed polymer chains—have also become possible by the SFA technique. SFA measurements for such steric repulsion have been made by several researchers (Klein, 1986; Malmsten *et al.*, 1990; Kurihara *et al.*, 1992; Kumacheva *et al.*, 1993). Figure 3.19 shows the surface force versus distance curves of the two mica surfaces bearing the brushlike adsorbed molecules of modified polymethacrylic acid (Kurihara *et al.*, 1992). One can see several interesting results from this figure on the steric interactions of the polymers: (1) The repulsive force extends up to submicrometer; (2) The strength and the extending length of the repulsive force are both very much dependent on pH in

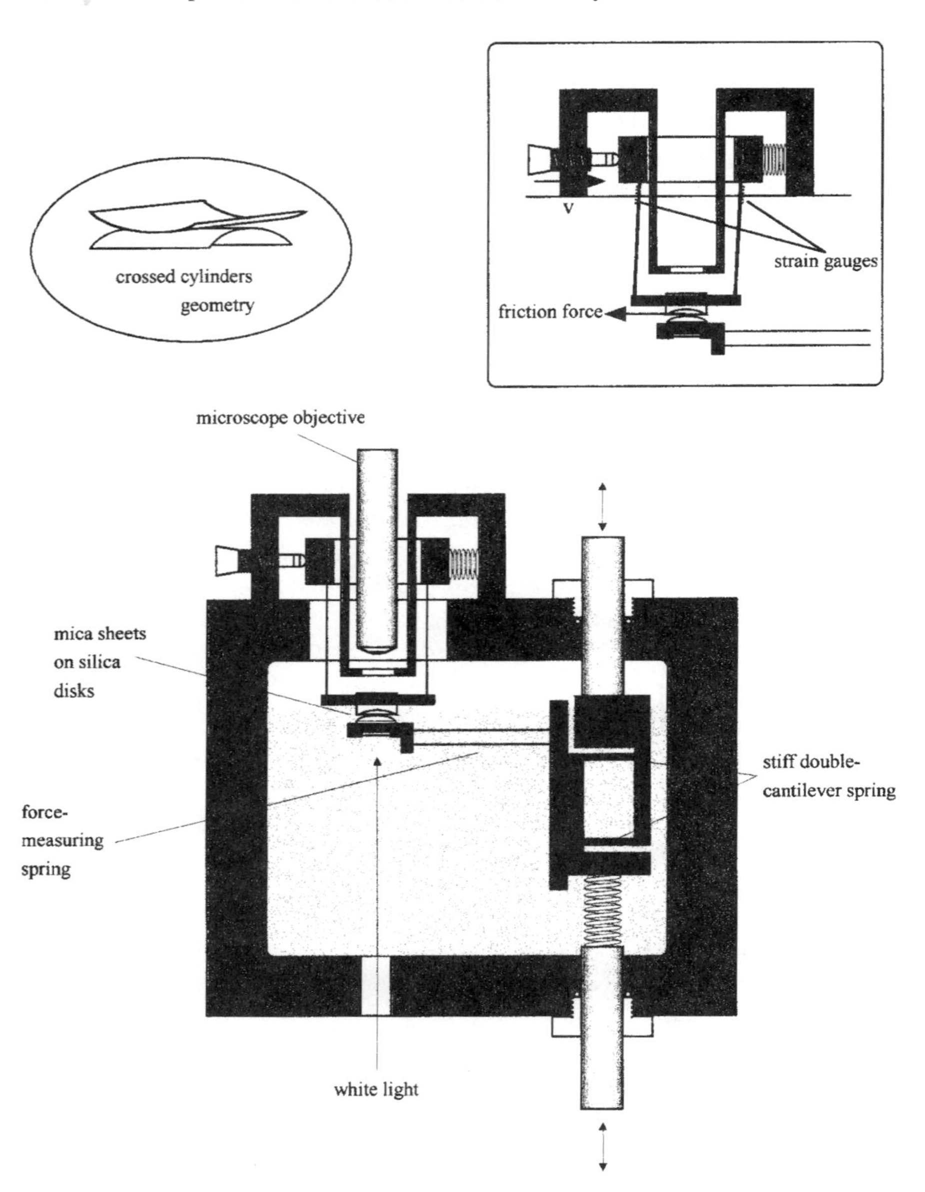

Figure 3.17 Surface force apparatus (SFA). Both the force acting between two mica surfaces and the frictional force when the mica surfaces are slid laterally can be measured by the SFA. The mica surfaces are set under the crossed cylinders geometry.

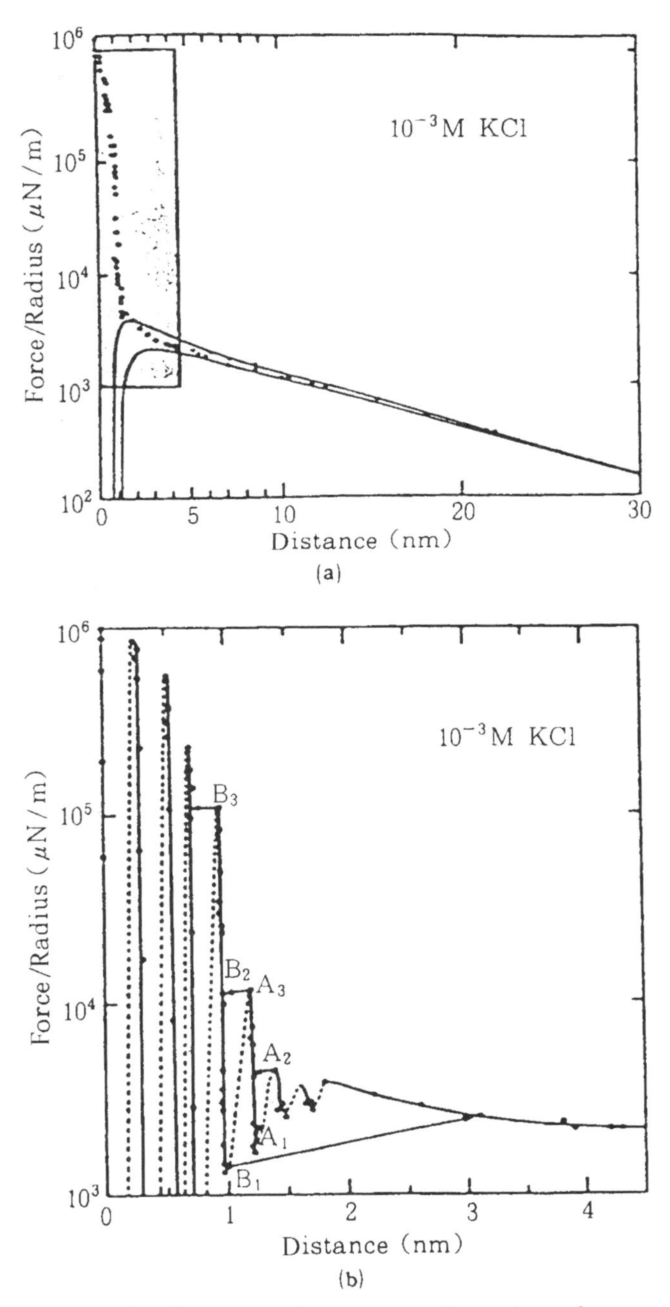

Figure 3.18 The surface force versus distance curve of two mica surfaces measured in an aqueous solution of 1.0 mM KCl (a), and its expansion in the region of a very short distance (b) (Pashley and Israelachvili, 1984). The surface force starts to oscillate when the two surfaces come closer to each other than 2 nm.

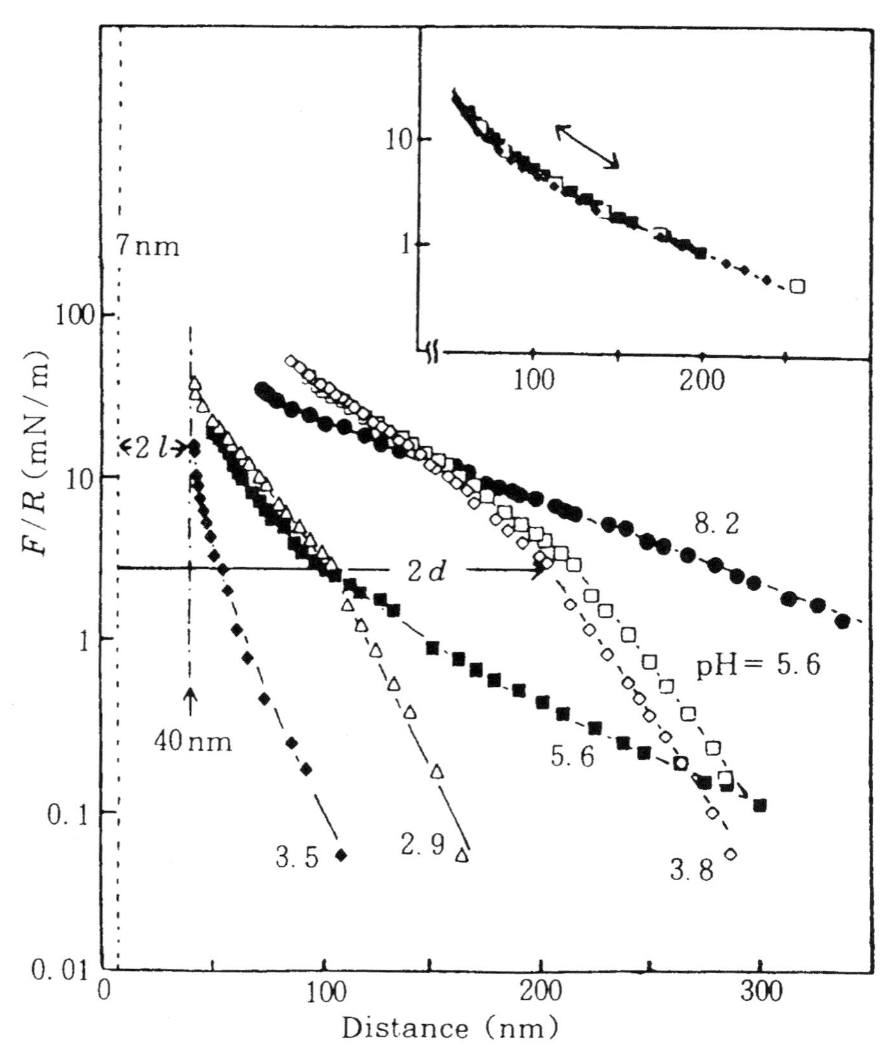

Figure 3.19 The surface force versus distance curves of the two mica surfaces bearing brushlike adsorbed polymer molecules of the modified polymethacrylic acid (Kurihara *et al.*, 1992). The filled and open symbols are data in pure water and in 100 mM NaBr, respectively. The pH values are denoted in the figure for each curve. The insert shows the reproducibility of the measurements. Reprinted with permission from *Langmuir,* **8,** 2087 (1992), American Chemical Society.

salt-free solutions; and (3) Two kinds of repulsion force are observed in the presence of salt since two slopes are observed in each force versus distance curve. It is reasonably accepted that the long-range repulsive interaction is mainly electrostatic and the short-range one is steric in the charged polyelectrolyte interactions. In fact, the salt effect on the interaction is much greater for the long-range repulsion. The nonionic surfactant of polyoxyethylene type is the most popular emulsifier, and the SFA measurement of the interaction force between two surfaces bearing this surfactant molecules is very informative. It is quite reasonable, and important, to see that the interaction is repulsive when the temperature is below cloud point (see Section 3.3.2.*b*) and attractive when the temperature is above it (Claesson *et al.*, 1986).

One of the most interesting results observed from the SFA may be the long-range attractive force between the hydrophobic surfaces in water (Israelachvili and Pashley, 1982; Pashley *et al.*, 1985; Kurihara and Kunitake, 1992). Figure 3.20

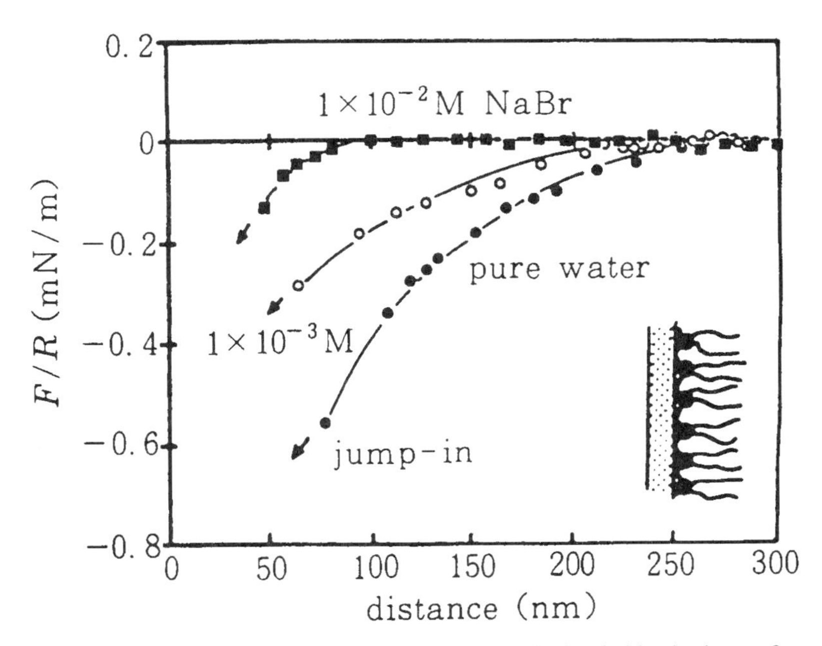

Figure 3.20 Long-range attractive force between two hydrophobized mica surfaces in pure water and aqueous NaBr solutions (Kurihara and Kunitake, 1992). The mica surfaces are coated with a polymerized double-chain cationic surfactant by the LB technique. Reprinted with permission from *J. Am. Chem. Soc.*, **114,** 10927 (1992), American Chemical Society.

shows some examples of these long-range attractive interactions (Kurihara and Kunitake, 1992). Note that the "hydrophobic" attractive force appears at a distance more than 200 nm in pure water. The normal hydrophobic interaction is believed to originate from the stronger cohesive energy between water molecules than that between water and hydrocarbons (see Section 1.3.1). So one may reasonably expect that the hydrophobic interaction must be in the short range of molecular order. Thus, it is still a mystery why this long-range attractive force is observed between the two hydrophobic surfaces.

A modified SFA technique has opened yet another new research field regarding surface forces: "nano-rheology." In this technique, two mica surfaces are slid laterally, and the friction is observed between the two surfaces having a separation distance of nm order (Israelachvili *et al.*, 1988; Gee *et al.*, 1990; Yoshizawa *et al.*, 1993). This observation is directly correlated with the boundary lubrication phenomenon. For example, Figure 3.21 shows the frictional force between two mica surfaces dipped in the liquid octamethyl-cyclotetrasilane (OMCTS; Gee *et al.*, 1990). The number n denoted in the figure indicates the number of molecular layers of OMCTS sandwiched between two mica surfaces. One can see in the figure that a molecular layer of OMCTS peels off during the friction measurement and that there is a stepwise increase in the friction force when one

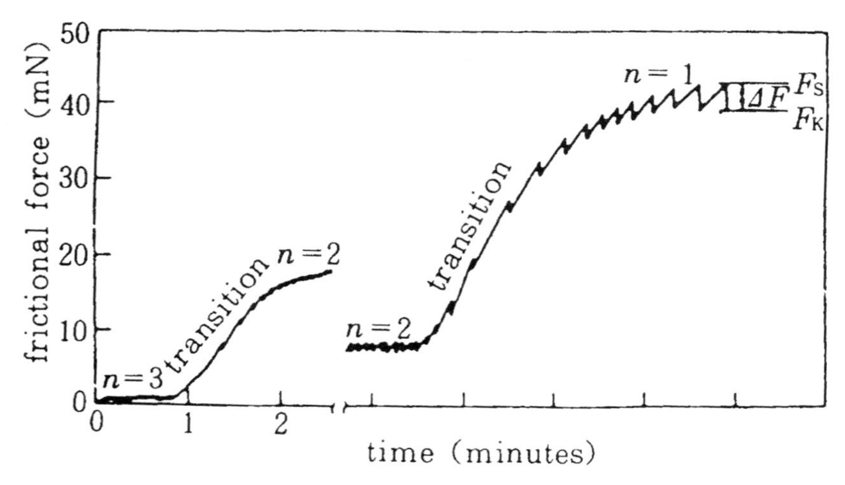

Figure 3.21 Frictional force between two mica surfaces dipped in liquid OMCTS, observed by the SFA technique (Gee *et al.*, 1990). The number n is the number of molecular layers of OMCTS between the mica surfaces.

molecular layer vanishes. The friction force under the presence of one molecular layer vibrates like a shape of saw teeth, which is called *stick-slip motion*. The stick-slip motion depends highly on the sliding velocity, as shown in Figure 3.22. One can see from the figure that the vibration frequency in the stick-slip motion increases with increasing sliding velocity. The stick-slip motion is thought to appear due to the repeated "phase transition" of OMCTS molecules between a solid (stick) state and a liquid (slip motion) state in the extremely thin space of two mica surfaces.

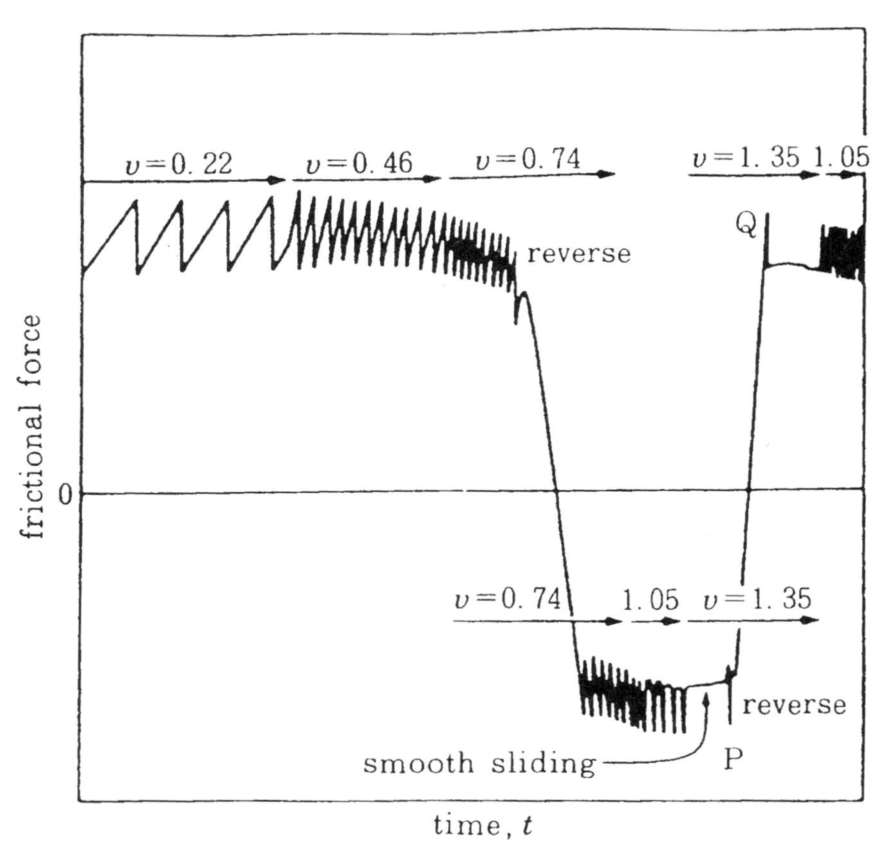

Figure 3.22 Stick-slip motion in two mica surfaces dipped in liquid OMCTS, observed by the SFA technique (Gee *et al.*, 1990). The vibration frequency of the stick-slip motion increases with increasing sliding velocity.

Another important contribution of the SFA technique to the tribology is to make clear the relationship between adhesion and friction forces (Yoshizawa and Israelachvili, 1994; Israelachvili *et al.*, 1994). People tend to consider that the adhesion force is directly correlated with the friction force. But this is the case only for static friction; kinetic friction is related to the hysteresis of a loading–unloading cycle of adhesion force. The hysteresis of adhesion force in a loading–unloading cycle results, of course, in the dissipation of energy during the process. The frictional motion is also a typical energy-dissipative process, so it is quite reasonable that it is related to the hysteresis of the adhesion cycle. For example, suppose that at the boundary between two liquidlike polymer surfaces the polymer molecules are intermixed with each other in part at the boundary, depending on the contact time of the two surfaces. When the loading–unloading cycle is slow, the intermixing of the polymer molecules proceeds well and the hysteresis in the adhesion force should be large. This molecular mechanism can be exactly applied to the kinetic friction process between the above polymer boundary.

e. Order-Disorder Phase Equilibria in Colloidal Systems

Order-disorder phase equilibrium or transition is one of the most important and main research targets in materials science of condensed matter. It is a phenomenon, of course, at the atomic or molecular level. A similar and possibly even more interesting order-disorder phase separation phenomenon occurs in colloidal systems. Monodispersed polymer latexes bearing high electric charges show a curious behavior when they are dialyzed after they are synthesized. Iridescent colored granules are separated out in milky white latex suspensions. This iridescent granule is a kind of colloid crystal and consists of ordered latex particles (e.g., Kose *et al.*, 1973). The colloid crystal diffracts the visible light (as the real (atomic) crystal does the X-ray) and shows the iridescence. The volume of the iridescent crystal phase increases with proceeding desalination by dialysis.

There are two theories at present to explain the above phase separation phenomenon of colloidal systems. If we apply the DLVO theory to the above latex system, only repulsive interaction between the particles is obtained, and we have to explain the phase separation in terms of the repulsive interaction only. If we adopt the theory of attractive interaction between particles, we have to abandon the DLVO theory (Ise, 1986; Sogami and Ise, 1984). Thus, both theories are still in the contest. Alder transition has been taken by Hachisu *et al.* in order to interpret the phase separation by only repulsive interaction (Kose and Hachisu, 1974; Hachisu and Kobayashi, 1974). The ordering of part of the particles allows more space for the remainder to gain the configurational entropy (Alder and Wainright, 1962; Alder *et al.*, 1968). This idea is essentially the same as that of Onsager's theory describing the phase separation of the needle-shaped tobacco mosaic virus (Onsager, 1942; Onsager, 1949).

One more new example of order–disorder phase separation has been added very recently—a system of plate-shaped bilayer membranes in surfactant gel phase (Yamamoto *et al.*, 1996). This novel iridescent phenomenon of the surfactant gel phase will be discussed in Section 3.3.3.*b*; only the TEM photographs of order and disorder bilayer membrane systems are given here in Figure 3.23.

f. Flocculation/Coagulation

Just as stability of dispersions is a kinetic phenomenon, so too is flocculation and/or coagulation. The terms *flocculation* and *coagulation* are not strictly discriminated, but flocculation is often used to describe a relatively weak or loose aggregation of particles that can be easily redispersed by a gentle shaking. Here we will use the term *coagulation* for the direct contact of two colloidal particles, falling down into the deepest ravine of the potential curve.

Dispersions are stabilized kinetically by the potential barrier shown in Figure 3.15, and coagulation should occur by reducing or extinguishing the barrier. When the potential barrier originates from electrostatic repulsion force, coagulation can be derived by eliminating the electric charges or compressing the Debye length. There is no clear means of depressing the potential barrier of the steric repulsion of adsorbed polymers; poor solvent for the adsorbed polymer may contribute to this.

Most colloid particles carry negative charges in aqueous solutions. Thus, cationic polymers and multivalent metal ions are useful as coagulation agents to eliminate the negative charges of the colloids. Alum and aluminum sulfate are frequently used since they provide trivalent aluminum ions. Added salt compresses the Debye length and reduces the electrostatic potential barrier, as demonstrated in Figure 3.24. Multivalent cations are also powerful for this shielding effect of electrostatic repulsion.

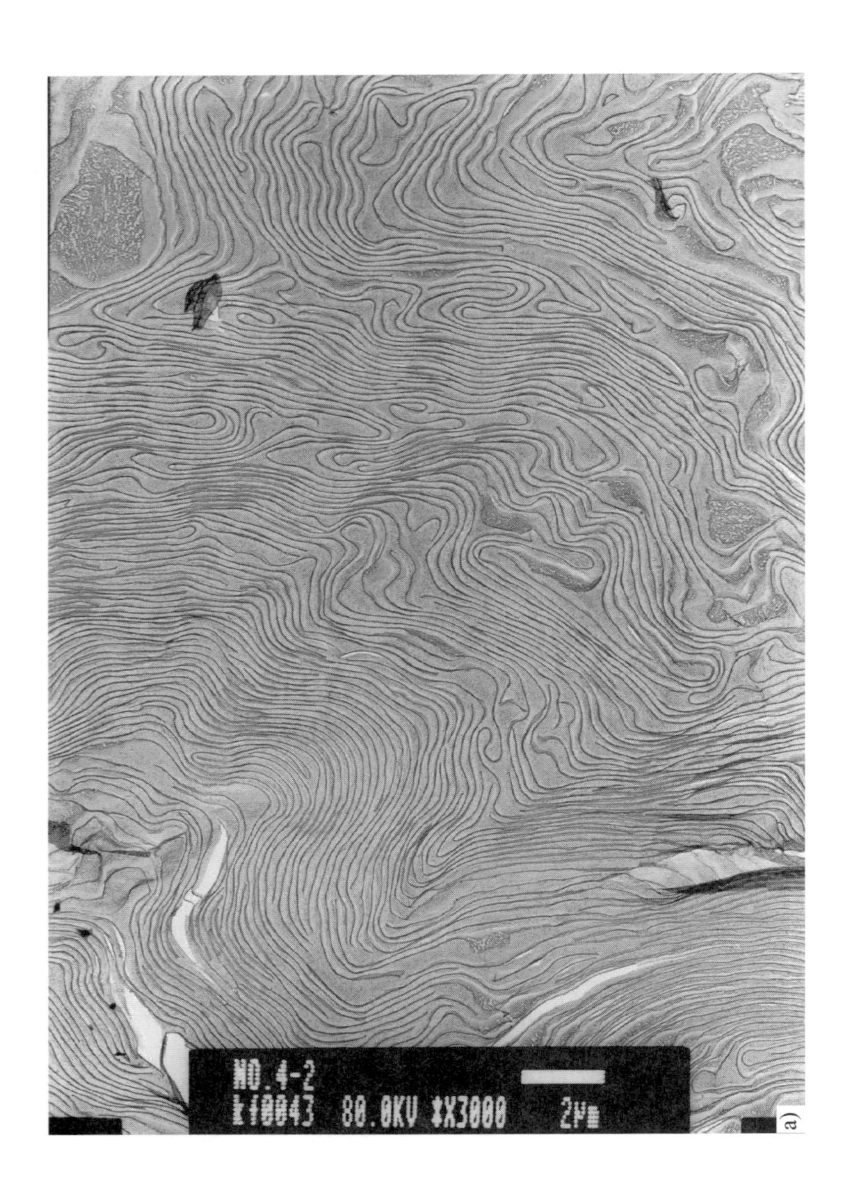
NO.4-2
kf0043 80.0KV *X3000 2μm
a)

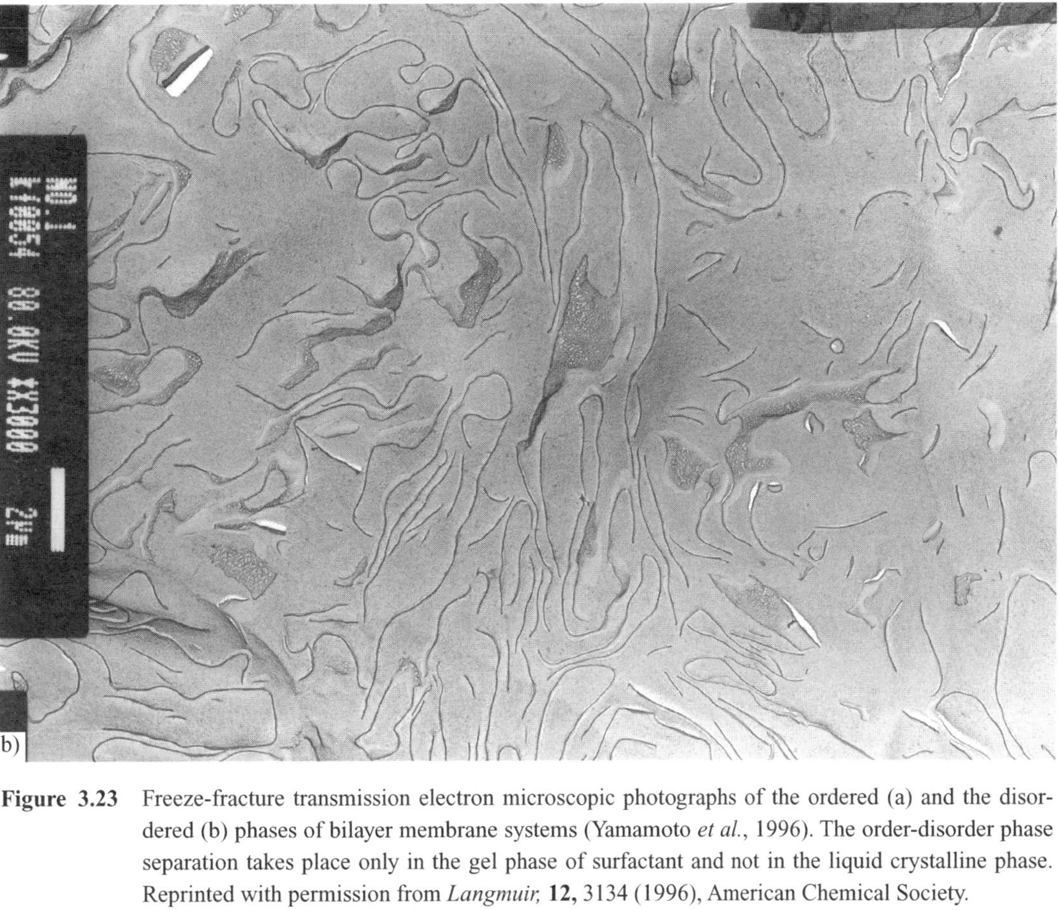

Figure 3.23 Freeze-fracture transmission electron microscopic photographs of the ordered (a) and the disordered (b) phases of bilayer membrane systems (Yamamoto *et al.*, 1996). The order-disorder phase separation takes place only in the gel phase of surfactant and not in the liquid crystalline phase. Reprinted with permission from *Langmuir,* **12,** 3134 (1996), American Chemical Society.

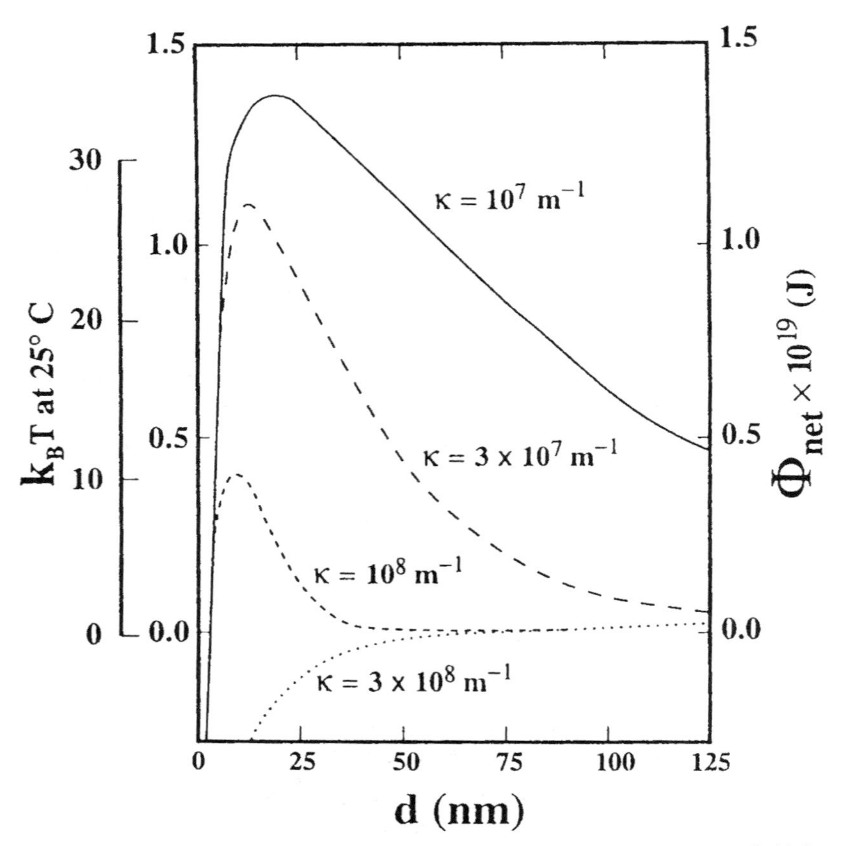

Figure 3.24 Effect of added salt on the total potential energy between two colloidal particles (Hiemenz and Rajagopalan, 1997). The potential barrier is reduced with increasing salt concentrations. The salt concentrations are expressed as κ (see Section 3.2.4.*b*). Reprinted from "Principles of Colloid and Surface Chemistry," 3rd ed., p. 587 by courtesy of Marcel Dekker, Inc.

3.3 Aggregation of Surface Active Substances and Related Phenomena

Hydrophobic chains of surfactant molecules aggregate with each other in aqueous solutions to avoid the contact with water after their adsorption is saturated on the available surfaces. Aggregation of two or three molecules of surfactant is not enough to effectively cut off the contact of hydrophobic groups with water mole-

Refreshing Room!

VERTICAL COMPONENT IN YOUNG'S EQUATION

Young's equation (3.2) is derived from the balance of horizontal components of two surface tensions and one interfacial tension (see Section 3.2.1.*c*). Some readers may be curious about the vertical component of the surface tension of the liquid. In Figure 3.6(a), the vertical component appears to be in imbalance with no counter force.

Counterbalance in both the horizontal and vertical components is clearly seen in the case of liquid–liquid contact (a lens; see Figure 3.25). The equilibrium shape should be like this even on the solid surface. The solid surface is, of course, very stiff and not deformed easily. So, Young's equation holds in a nonequilibrium situation. But the liquid shape is stable because the vertical component is counterbalanced with an elastic force of the solid. The vertical component of the liquid surface tension pulls the solid surface at the contact line. The solid surface copes elastically with this force without deformation of its shape.

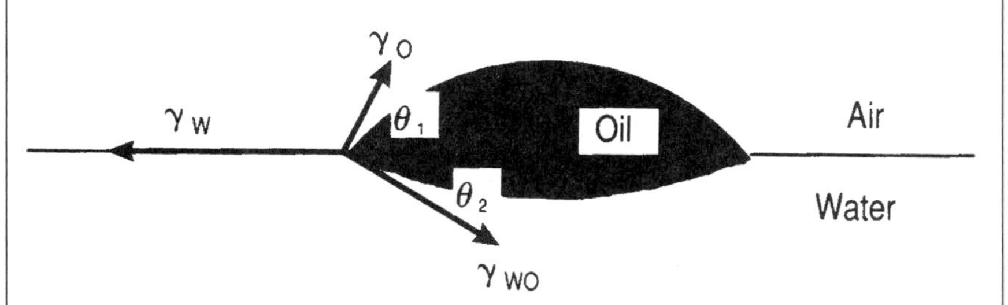

Figure 3.25 Equilibrium shape of a liquid on another liquid (a lens). The balance of two surface tensions and one interfacial tension holds in both the horizontal and vertical components in this case, i.e., $\gamma_W = \gamma_O \cos \theta_1 + \gamma_{WO} \cos \theta_2$ and $\gamma_O \sin \theta_1 = \gamma_{OW} \sin \theta_2$.

cules. So several tens of molecules (or even more) gather together and form micelles, liquid crystals, vesicles (liposomes), and so on. The term *micelle* is used for an individual body of aggregate having a globular, cylindrical, or platelike shape. Liquid crystals are formed in a higher concentration region resulting from the periodic arrangement of the micelles in the long-range order. Lamellar liquid crystals formed with the platelike micelles are multilayered vesicles having an onion-like structure when they are separated out of and dispersed in water medium. The single bilayer vesicle is a special case of lamellar liquid crystal that has only one bilayer membrane.

3.3.1 PHYSICAL SIGNIFICANCE OF THE KRAFFT POINT

a. What is the Krafft Point?

A surfactant shows a very unique behavior in its solubility-temperature curve. When we measure the solubility of a surfactant into water, we may find a critical temperature at which the solubility increases abruptly. This critical temperature is called the *Krafft point* or *Krafft temperature* after the discoverer of this phenomenon (Krafft and Wiglow, 1895; Krafft, 1896; Krafft, 1899). Solubility of the surfactant is quite low below the Krafft point; the surfactant is practically insoluble into water. Consequently, surface activity cannot be obtained below this temperature. Thus, the Krafft point is one of the fundamental characteristic physical quantities of surfactants, together with the critical micelle (micellization) concentration (CMC) (described later in detail).

The physical meaning of the Krafft point can be understood from the phase diagram of a dilute aqueous solution of a surfactant shown in Figure 3.26. Curve BAC, the apparent solubility curve, suddenly increases at the Krafft point. Curve AD is the CMC curve at which the micelle is formed when the surfactant concentration is increased above the Krafft temperature. These two curves intersect at point A, at which the surfactant solubility starts to increase. The surfactant is in solid (hydrated crystal) state in region (I) when added over the solubility limit and dissolves as a monomer (molecularly dispersive) solution in region (III). In region (II), micelles are equilibrated with monomers of saturated concentration (CMC). The solubility-temperature curve of any material must vary smoothly, like curve BAE in Figure 3.26, if the state of the material does not change. If the material melts with elevating temperature, the solubility curve must break, like curve BAD, because the temperature-dependence of solubility becomes smaller in a liquid state. The solution enthalpy that determines the slope of the solubility-temperature curve becomes smaller by the enthalpy of fusion of the material. Accordingly, the solubility curve must be like BAE or BAD, and never like BAC.

Let us consider a phase diagram of an ordinary substance, say, benzene, to overcome the above contradiction between the theoretical and the apparent solubility curve of a surfactant. The solubility curve of solid benzene may be like curve BA and break at its melting point like curve BAD. In this situation solid

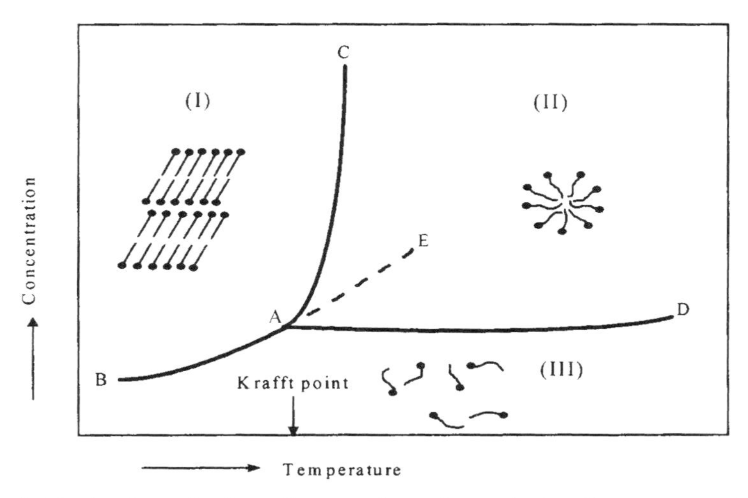

Figure 3.26 A schematic phase diagram of a surfactant in the dilute concentration region of an aqueous solution. Apparent solubility of the surfactant increases abruptly at the Krafft point.

and liquid benzene are separated out of the aqueous solution in regions (I) and (II), respectively. Thus, the boundary AC is the melting temperature of benzene. According to this interpretation, micelles are the liquid phase of a surfactant separated out of the aqueous solution, and the Krafft point is the melting temperature of the solid (hydrated crystal) state of a surfactant (the phase separation model of micelles, Shinoda and Hutchinson, 1962; Shinoda, 1963). Solution must separate into two bulk phases or become turbid when the phase separation of liquid benzene occurs. But in the case of surfactant solutions, the two-phase region is still transparent because of very small liquid droplets (micelles) having a diameter of ~5 nm. This is why the apparent solubility curve of a surfactant behaves abnormally like curve BAC.

b. Phase Transitions Between Solid Phases in Surfactant/Water Systems

Since most are interested in solutions of surfactants, little attention has been paid to the solid state of the agents. However, one may find some interesting solid phases and phase transitions even below the Krafft point, and some of them are practically applied. Figure 3.27 shows the phase diagram of an octadecyltrimethylammonium chloride/water system observed by a differential scanning calorimetric technique (Kodama and Seki, 1991). Note that the ordinate and

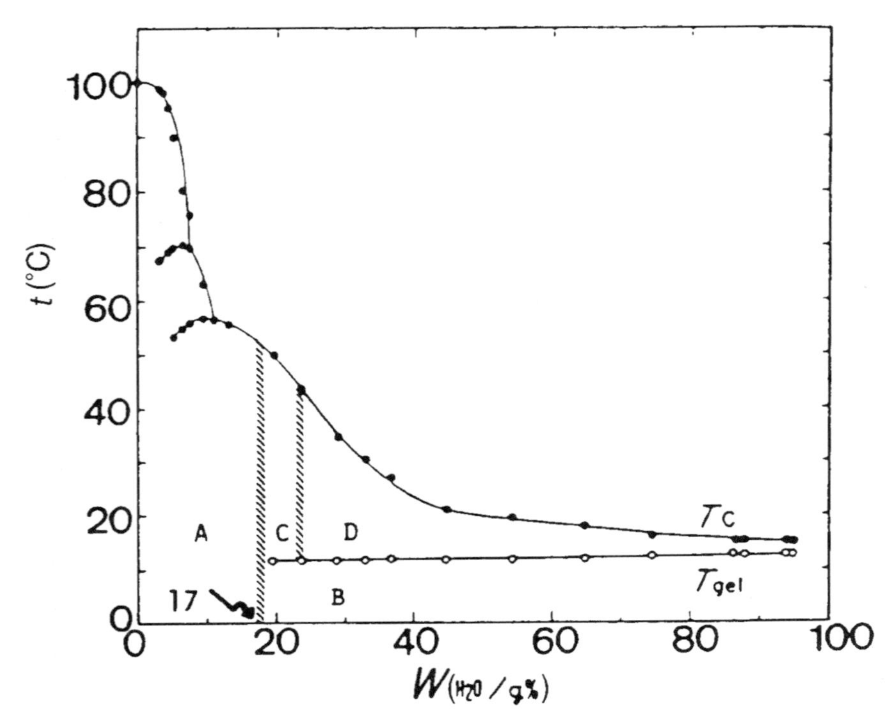

Figure 3.27 Phase diagram of the octadecyltrimethylammonium chloride/water system. A solid-solid (coagel-gel) phase transition is present below T_C. Reprinted from *Adv. Colloid Interface Sci.,* **35,** M. Kodama and S. Seki, p. 1 (1991) with kind permission of Elsevier Science-NL, Sara Burgerhartstraat 25, 1055 KV Amsterdam, The Netherlands.

the abscissa are exchanged in this figure, and the concentration range is much wider than in Figure 3.26. CMC cannot be drawn in this figure since the CMC curve overlaps on the ordinate of the right side. The curve denoted as T_C is called the *gel–liquid crystalline phase transition* curve and is the Krafft point curve itself. The curve T_{gel}, found below T_C, is known as the *coagel-gel phase transition* curve. The coagel is a true (hydrated) crystalline phase, and the gel phase consists of two-dimensional crystals of the bimolecular leaflets and the water layers between them. Molecular models for these phases as well as for the liquid crystalline phase are illustrated schematically in Figure 3.28. In the coagel (crystalline) phase, surfactant molecules are ordered in both positional and orientational freedoms, but they are in orientational disorder in the gel phase. The

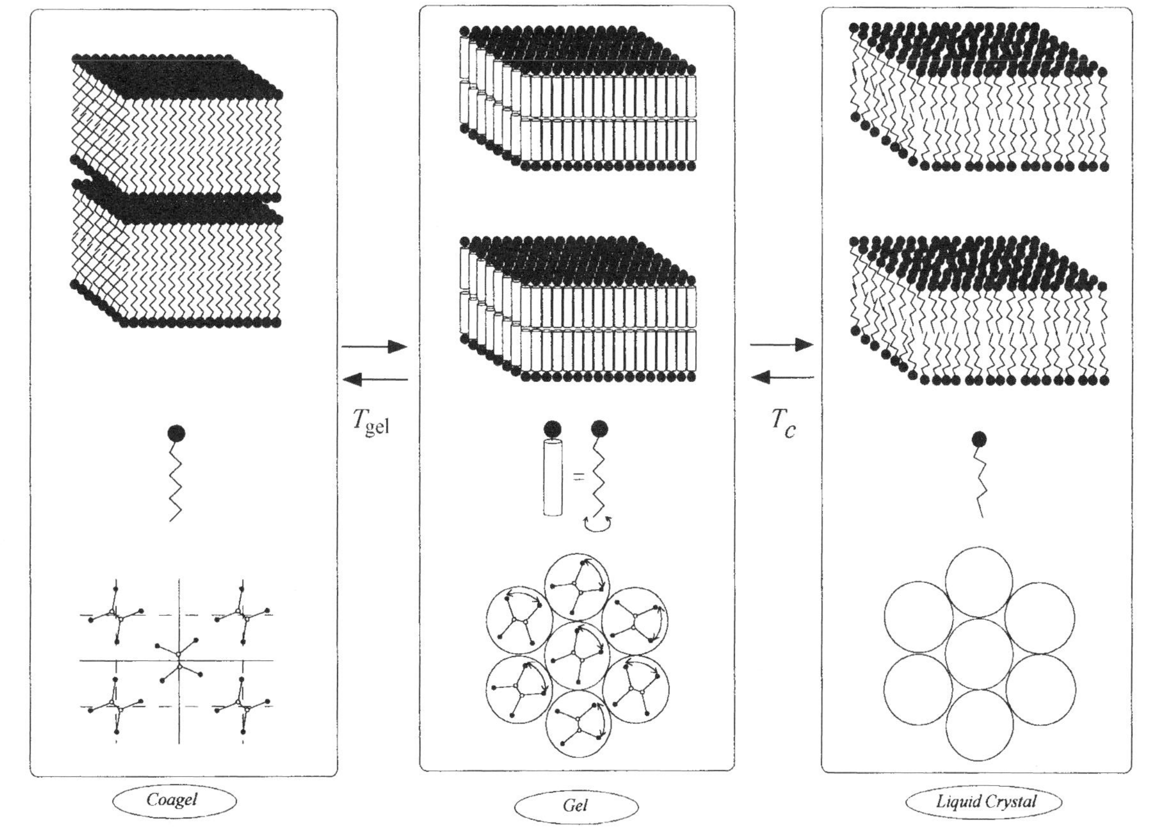

Figure 3.28 Schematic illustration of the molecular arrangement of the coagel, gel, and liquid crystalline phases of a surfactant. The surfactant molecules possess their positional and orientational order in the coagel phase, but the orientational order is lost in the gel phase. In the liquid crystalline phase, the molecules are in a random state and the hydrocarbon chains move flexibly.

surfactant molecules are disordered in both position and orientation, and hydrocarbon chains are in flexible movement in the liquid crystalline phase.

The coagel-gel phase transition is essentially the order-disorder transition on the rotational freedom of surfactant molecules. But quite interestingly, the distance between bilayer membranes dramatically increases from the order of 1 to 100 nm at this transition point simultaneously with the rotational transition of the molecules. The rotational transition causes an increase in the area occupied by one surfactant molecule in bilayer membranes and a decrease in the surface charge density. The lower charge density at the surface of the bilayer membranes results in a higher degree of dissociation and a more homogeneous distribution of counterions in the aqueous phase between the membranes. Both effects contribute to the stronger repulsive force between the bilayers, and the membrane separation takes place (Tsuchiya *et al.*, 1994).

c. Krafft Points of Mixed Surfactant Systems

The melting point model of the Krafft point is quite useful for practical applications of surfactants. This model provides insight into how to depress the Krafft point, which is important for practical applications of surfactants since the agents do not show any surface activity below the Krafft point. Krafft point depression, similar to the melting point depression phenomenon in ordinary substances, is also observed in mixed surfactant systems. Figure 3.29 shows the

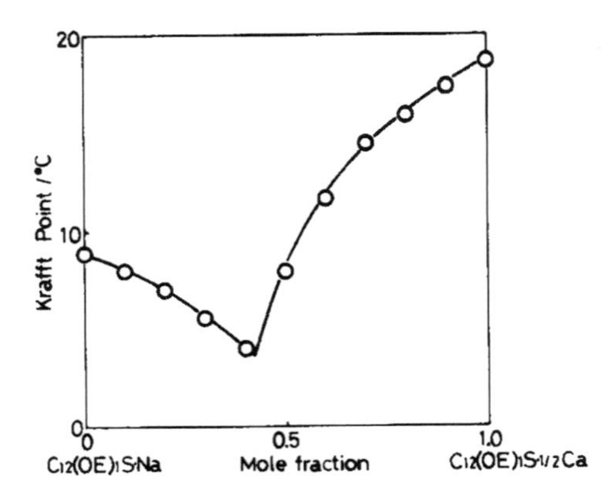

Figure 3.29 Krafft point-composition curve of the binary mixture of sodium and calcium dodecyloxyethylene sulfate (Tsujii *et al.*, 1980). Reprinted with permission from *J. Phys. Chem.,* **84,** 2287 (1980), American Chemical Society.

Krafft point-composition curve in a binary mixture of sodium and calcium dodecyl-oxyethylene sulfate (Tsujii *et al.*, 1980). The Krafft points of the mixed surfactant system are lower than those of either pure component, which shows the melting (Krafft) point depression phenomenon. The melting point depression phenomenon takes place when both components are mixed in the liquid (micelle in this case) phase and not mixed in the solid phase. The quantitative expression of the melting point depression is

$$-\ln X_A = \frac{\Delta H_A^0}{R}\left(\frac{1}{T} - \frac{1}{T^0}\right), \tag{3.16}$$

where X_A is the mole fraction of surfactant A in the micellar phase, ΔH_A^0 and T^0 are the enthalpy of fusion and the Krafft temperature, respectively, of pure surfactant of component A, T is the Krafft point of the mixed system, and R is the gas constant. Let us emphasize again that X_A is the mole fraction of component A *in the mixed micelles.* We are dealing with the solutions of two kinds of surfactant (the mixed micelles), not the solutions of water. Water is regarded as just a vessel of the solutions of surfactants since the micelle is the separated phase of the aqueous solution. The values of ΔH_A^0 and ΔH_B^0 were calculated from the data of Figure 3.29 using Eq. (3.16), and were in good agreement with those observed experimentally.

When two components are mixed in the liquid phase and not mixed in the solid phase, the melting point depression phenomenon takes place and an eutectic point appears. The immiscibility of surfactant components in the solid phase has been substantiated in eutectic mixtures of 3-hydroxy-1-pentadecane sulfonate/2-pentadecene-1-sulfonate (Figure 3.30, Tsujii *et al.*, 1980). A Krafft point versus composition curve of the system displaying complete miscibility in both the liquid and solid phases is shown in Figure 3.31. In this case, the Krafft points change monotonously. Composition analysis of solid samples actually shows the complete miscibility (solid solution) of this system.

Figure 3.32 shows one more type of the Krafft point versus composition curves obtained in a surfactant mixture of strong attractive interaction. Specifically, the figure shows the Krafft point versus composition curves of binary mixtures of sodium dodecyl sulfate and amphoteric alkylsulfobetaine (Tsujii *et al.*, 1982). Addition compounds between the anionic and amphoteric surfactant of 1/1 or 1/2 molar ratio are formed in these mixed systems. Similar addition compound formation was found in other combinations of anionic and amphoteric surfactants. In each mixture, two eutectic points appear between the addition compound and anionic surfactant as well as between the addition compound and amphoteric surfactant. Addition compound is formed probably due

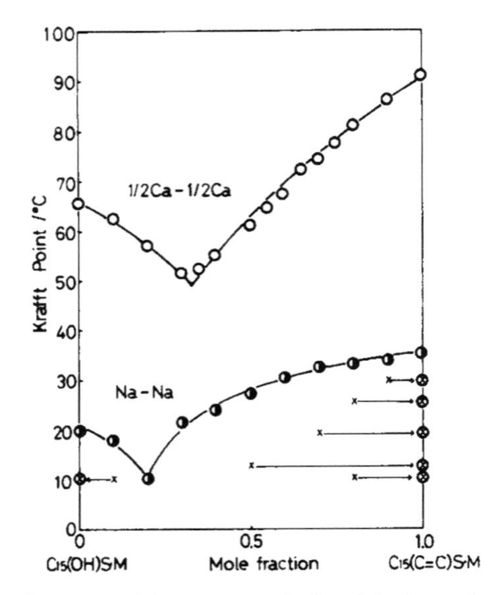

Figure 3.30 Krafft point-composition curve of the 3-hydroxy-1-pentadecane sulfonate/2-pentadecene-1-sulfonate system (Tsujii *et al.*, 1980). The ⊗ marks indicate the composition of the solid phase collected at the given temperature and mixing ratio marked with × signs. Reprinted with permission from *J. Phys. Chem.*, **84,** 2287 (1980), American Chemical Society.

to the attractive interaction between anionic sulfate and cationic quarternary ammonium ions.

d. How to Depress the Krafft Point

As mentioned previously, it is important in practical applications of surfactants to depress the Krafft point. The following methods for depressing the Krafft point can be predicted from the melting point model.

1. *Insert the branched structure and/or unsaturated bonds in the hydrophobic chain of surfactant molecules.* This method is easily expected from the low melting point of branched and/or unsaturated hydrocarbons.
2. *Alter the counterions of the ionic surfactant.* Counterion alteration modifies the crystal structure of surfactant, thus changing the Krafft

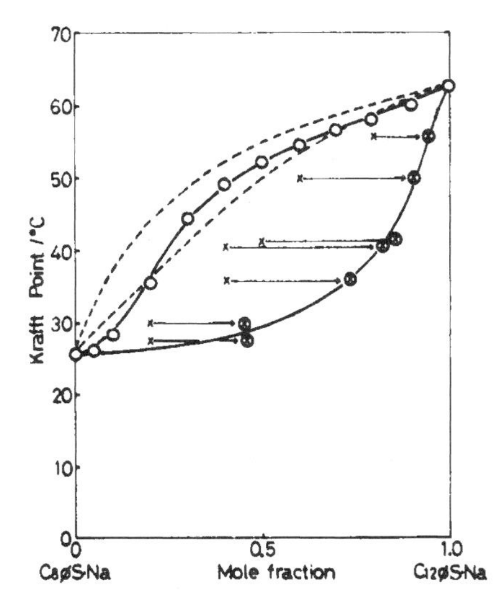

Figure 3.31 Krafft point-composition curve of the sodium octylbenzene sulfonate/sodium dodecylbenzene sulfonate system (Tsujii *et al.*, 1980). The Krafft point increases monotonously in this case. The signs of ⊗ and × have the same meaning as those in Figure 3.30. The theoretical curve calculated under the assumption of ideal mixing is shown by dotted lines. Reprinted with permission from *J. Phys. Chem.*, **84,** 2287 (1980), American Chemical Society.

point. It is difficult, however, to predict what kind of counterion lowers the Krafft point. Triethanolammonium is empirically well known as a counterion for anionic surfactants to obtain a low Krafft point.

3. *Insert the polyoxyethylene and/or polyoxypropylene group in the surfactant molecule.* The polyoxypropylene group is more effective since its melting point is lower than that of the polyoxyethylene group.
4. *Blend with some other surfactants.* This method was mentioned in the previous section and is an application of the melting point depression phenomenon.
5. *Add some organic compounds, which are incorporated (solubilized) into micelles.* The mole fraction of surfactant in the micellar phase is depressed by the addition of the organic substance. The melting point depression phenomenon is again utilized in this method. Fatty alcohols and sodium *p*-toluene sulfonate are often used in this method.

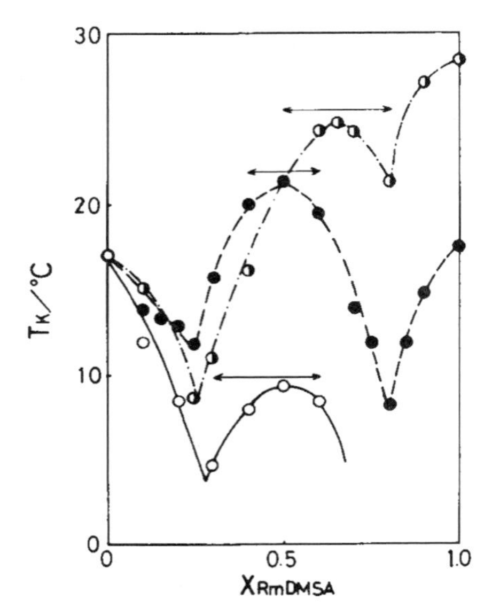

Figure 3.32 Krafft point-composition curves of sodium dodecyl sulfate/alkylsulfobetaine systems (Tsujii *et al.*, 1982). Alkyl chains of the sulfobetaines are C_{12} (○), C_{14} (●), and C_{16} (◑). Reprinted with permission from *J. Phys. Chem.*, **86,** 1437 (1982), American Chemical Society.

6. *Add inorganic salts.* This method is used only in the case of zwitter-ionic surfactants (Tsujii and Mino, 1978). Thiocyanate ion is particularly effective. It must be noted that the addition of salts gives the completely opposite effect on the Krafft point of ionic surfactants.

3.3.2 MICELLE FORMATION, SOLUBILIZATION, AND MICROEMULSIONS

a. Micelle as a Liquid Droplet

We mentioned in the previous section that the micelle is a separated liquid phase. Let us consider in more detail the phase separation model for micelles. Although we pointed out in Section 3.2.1.*a* that the Gibbs' adsorption isotherm cannot be applied to calculate the adsorption amount in the concentration range above CMC, we dare to do so here to get some insight into the physical significance of the nature of micelles. Surface tension is almost constant at concentrations higher than CMC, i.e., $d\gamma = 0$. The adsorption amount, Γ, on the left-hand side of Eq. (3.1) has been substantiated experimentally to be not zero. Consequently, the necessary condition to hold Eq. (3.1) is $d \ln C = 0$, i.e., C = constant. Surfactant concentration does not increase even when the agent is added to the solution, which means that the phase separation of the surfactant takes place. This is a thermodynamic proof of the phase separation model of surfactant micelles.

There is also much evidence to substantiate the liquid state of micelles. The partial molar volume of micelles is close to that of liquid hydrocarbons (Shinoda and Soda, 1963). NMR studies themselves on micellar solutions give a strong indication of the liquid nature of micelles because the solutions give a beautiful high-resolution spectra in extreme narrowing conditions. In addition, one NMR study clearly showed that the tail end of the hydrocarbon chains of surfactants came close to the surface of the micelles (Tokiwa and Tsujii, 1972). These results indicate the flexible movement of hydrocarbon chains in the micelles. The following observations are also indirect verifications of the liquid state of micelles: (1) Two kinds of surfactant are readily miscible to form the mixed micelles and (2) Organic materials can be easily incorporated into micelles (solubilization phenomenon). The solid materials do not incorporate other compounds so easily, therefore we can purify the materials by recrystalization.

A micelle is an aggregate of several tens of surfactant molecules. Accordingly, most of the solution properties of the surfactant of course change drastically at CMC. This means that most of the solution properties can be used to determine the CMC. In fact, more than 20 methods have been given to determine the CMC (Shinoda, 1963; Mukerjee and Mysels, 1971), but the methods of surface tension, solubilization of water-insoluble dyes, and spectral change of water-soluble dyes are most common and most useful. To determine the CMC of ionic surfactants, the electric conductivity method is the most accurate (Goddard and Benson, 1957), but this method cannot be applied to solutions containing added electrolytes.

b. Micellar Size and Shape

A micelle consists of several tens of surfactant molecules or even more. The number of molecules in one micelle is the *aggregation number* of the micelle, and the aggregation number multiplied by the molecular weight of the surfactant is the *micellar* (molecular) *weight.* Table 3.2 shows the aggregation number and micellar weight of sodium dodecyl sulfate in pure water and aqueous NaCl solutions. The aggregation number is from several tens to one hundred and several tens, depending on the concentration of sodium chloride. These values are typi-

Table 3.2 Micellar Weight and Micellar Aggregation Number of Sodium Dodecyl Sulfate

Medium	*Micellar Weight*	*Aggregation Number*
Pure water	17,800	62
	23,200	80
	25,600	89
0.01 mol/l NaCl	25,600	89
0.02 mol/l NaCl	19,000	66
0.03 mol/l NaCl	23,500	72
	28,700	100
	29,500	102
0.05 mol/l NaCl	30,100	105
0.1 mol/l NaCl	31,600	110
	32,200	112
0.2 mol/l NaCl	29,500	101
0.5 mol/l NaCl	41,000	142

cal for hydrophilic water-soluble ionic surfactants. Micelles are usually spherical or globule, with a diameter of about 5 nm. The micellar size of nonionic surfactants exhibits a unique behavior with temperature elevation. The aggregation numbers of hexaoxyethylene dodecyl ether and undecyl carboxybetaine plotted against temperature are shown in Figure 3.33 (Tanford, 1980b). The aggregation number of the nonionic surfactant micelle increases steeply with temperature in contrast to that of the amphoteric agent. The aggregation of the nonionic surfactant further proceeds, and finally the liquid surfactant phase is separated out at a certain temperature. This temperature is called the *cloud point* since the solution gets turbid at this temperature. The clouding phenomenon indicates that the nonionic surfactant/water system has a lower critical solution temperature (LCST) type phase diagram, and the dissolution of the agent into water is enthalpy-driven. The hydration to ether oxygen of the polyoxyethylene chain is the driving force to dissolve the surfactant. Therefore, the polyoxyethylene chain of nonionic surfactant loses its hydrophilic nature by dehydration with increasing temperature, and the agent becomes water-insoluble. This is the mechanism of the clouding phenomenon of nonionic surfactants.

Micellar size is closely related to the shape of the micelles. When the aggregation number of the micelle is increased, the only way to keep its shape spherical is to enlarge the radius of the sphere and to make a vacancy in the core of the micelles. Therefore the shape of the micelle will be deformed from spherical be-

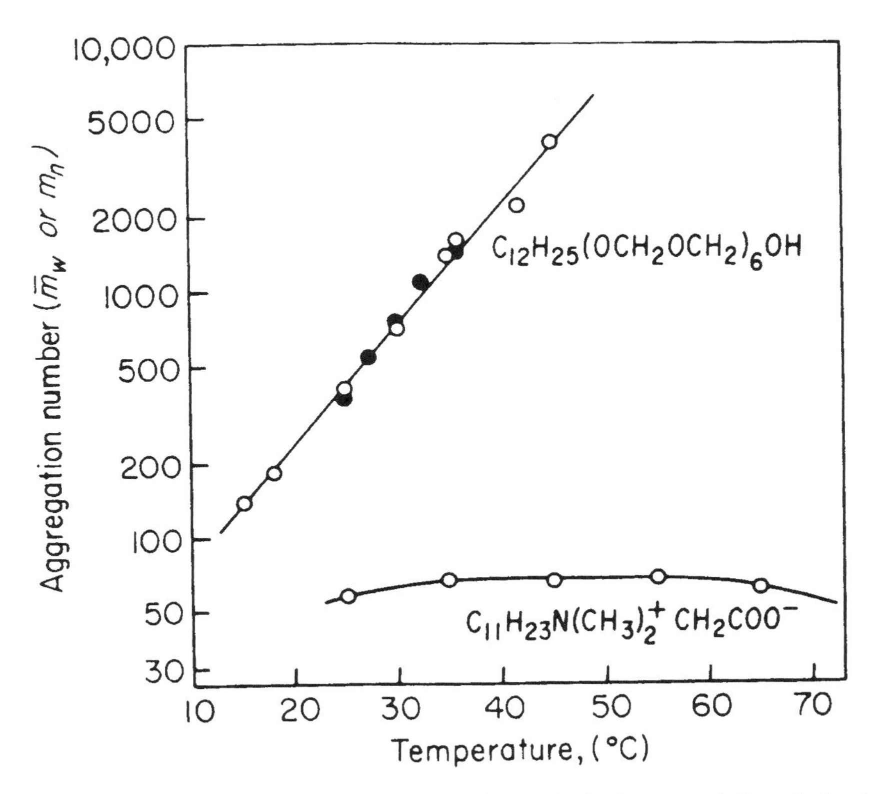

Figure 3.33 Micellar aggregation numbers of a nonionic (hexaoxyethylene dodecyl ether) and an amphoteric surfactant (undecyl carboxybetaine) as a function of temperature. "The Hydrophobic Effect" 2nd ed. by C. Tanford, copyright © 1980. Reprinted by permission of John Wiley & Sons, Inc.

cause vacancy formation inside the micelle is impossible. Let us consider here the factors that determine the shape of micelles. According to Tanford (Tanford, 1972; Tanford, 1980b) and Ninham's group (Israelachvili *et al.*, 1976; Israelachvili *et al.*, 1977; Mitchell and Ninham, 1981), the governing factor of the shape of a micelle is the packing of hydrophobic chains of surfactant molecules. Figure 3.34 illustrates this relationship between the packing of the hydrocarbon chains and the shape of micelles. As seen on the left-hand side of the figure, surfactant molecules with a small hydrophobic chain relative to the hydrophilic head tend to form spherical micelles when packed together without any vacancy inside the micelle. When the hydrophobic and hydrophilic groups are the same

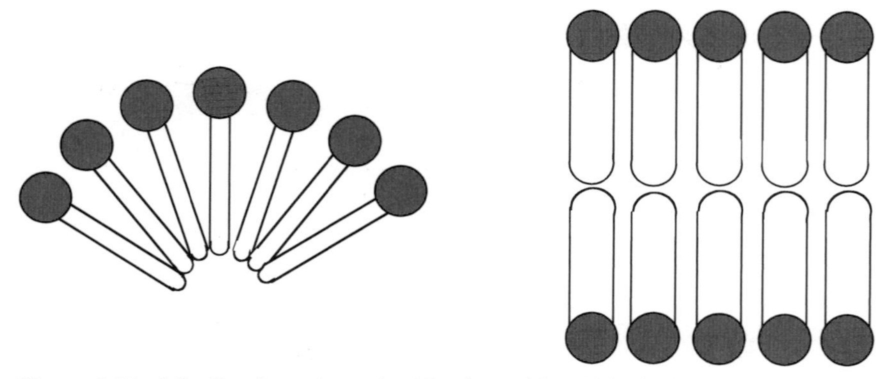

Figure 3.34 Micellar shape determined by the packing of the hydrocarbon chains of a surfactant

size, the micelle tends to be flat, as illustrated on the right-hand side of the figure. The size of hydrophilic head includes not only the atomic size of the head group but also the space resulting from the electric repulsion between the ionic heads of the surfactant molecules. This is why ionic surfactants readily form spherical micelles. Double-chain surfactants with big hydrophobic groups, on the other hand, form the platelike micelles to give vesicles and liposomes.

Figure 3.35 shows the quantitative explanation of the above packing concept for micellar shape. Supposing a spherical micelle consisting of n surfactant molecules with a hydrophobic tail of length l and of volume v, the following equations hold.

For the surface area of the micelle: $4\pi l^2 = ns$

For the volume of the micelle: $\frac{4\pi l^3}{3} = nv$

Here, s is the area occupied by one hydrophilic head of a surfactant molecule at the surface of the micelle. Eliminating n from the above two equations, we obtain $v/ls = 1/3$. Spherical micelles form more easily if s is larger than the above condition, and the following inequality is obtained as the condition of the spherical micelle:

$$\frac{v}{ls} \leq \frac{1}{3}. \qquad (3.17)$$

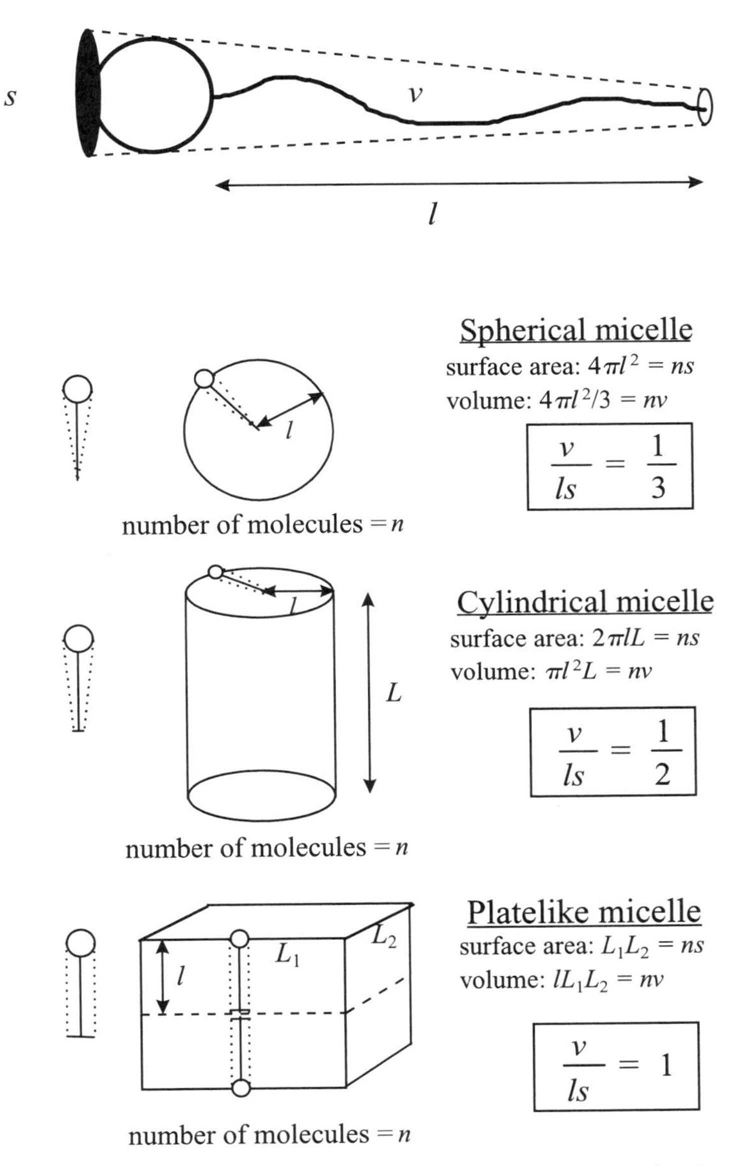

Figure 3.35 Quantitative representation of the packing theory to determine the micellar shape.

Similarly, the conditions for cylindrical (rodlike) and flat (platelike) micelles can be written as follows (see Figure 3.35).

$$\text{For cylindrical micelles: } \frac{1}{3} < \frac{v}{ls} \leq \frac{1}{2} \tag{3.18}$$

$$\text{For platelike micelles: } \frac{1}{2} < \frac{v}{ls} \leq 1 \tag{3.19}$$

In principle, we can predict the micellar shape from these inequalities when some surfactant molecule is given. It is not easy, however, to estimate the parameters of v, l, and s practically. The value of v may be estimated under some assumptions. The length l must be somewhat shorter than that of the extended hydrocarbon chain, but we do not know how much shorter. It is impossible to estimate the value of s because it is affected by too many factors—ionic repulsion, hydration, chemical structure of hydrophilic group, surfactant concentration, added salts, pH (in some cases), and so on. The packing theory is still useful, however, for understanding the change of micellar shape qualitatively for a given surfactant under changing conditions of the solutions.

Solution properties of a surfactant, of course, highly depend on the shape of micelles. Viscoelastic properties are particularly sensitive to it (Hoffmann and Ebert, 1988; Hoffmann, 1994). Long cylindrical micelles behave like polymers in solutions and give the solutions a viscoelastic nature. Entanglements of long threadlike molecules or aggregates are the main origin of elasticity of solutions. In real polymer solutions, entanglements can be loosened by only the reptation mechanism. In the surfactant solutions, we have several kinds of systems in which the entangled cylindrical micelles can loose themselves by different mechanisms. A remarkable characteristic of these systems is that the entangled two cylindrical micelles basically can pass through each other by rearrangement of their molecular assemblies. Much theoretical attention has been given to discover how this temporary entanglement behavior occurs. Platelike (flat) micelles also show an interesting viscoelastic nature having a yield value. The basic structure of this solution is multilamellar vesicles, which will be described in Section 3.3.4.*b*. Spherical vesicles are densely packed in the whole solution and give the elastic response to the deformations (Hoffmann, 1994).

c. Solubilization

A micelle can be regarded as a kind of very tiny liquid droplet. As a result, organic substances are able to dissolve into the micellar core of the tiny organic solvent. Thus, some organic compounds that are insoluble in water do dissolve into aqueous micellar solutions. This phenomenon is called *solubilization,* and the solubilized organic material is called the *solubilizate.* In this broader sense of solubilization, the organic material is not necessarily water-insoluble. We can also use the term *solubilization* when the solubility of some water-soluble compound increases dramatically in the presence of micelles. Note that the solubilization phenomenon occurs only when the micelles are present in the solutions, and never in the monomer solutions of surfactant. So we can use the solubilization phenomenon as a method of CMC determination.

As reasonably understood, the locations of solubilized molecules in the micelle are different from compound to compound. Figure 3.36 shows a schematic illustration of the solubilized positions of some organic substances. Nonpolar organic compounds are solubilized in the hydrocarbon core of micelles, whereas polar substances (such as higher alcohols) are in parallel with surfactant molecules, orientating their polar group toward the water medium. A condensed layer of polyoxyethylene chains is present around the hydrocarbon core of a nonionic surfactant micelle. Some of the organic materials (such as aromatic dyes) are preferably solubilized in this polyoxyethylene shell

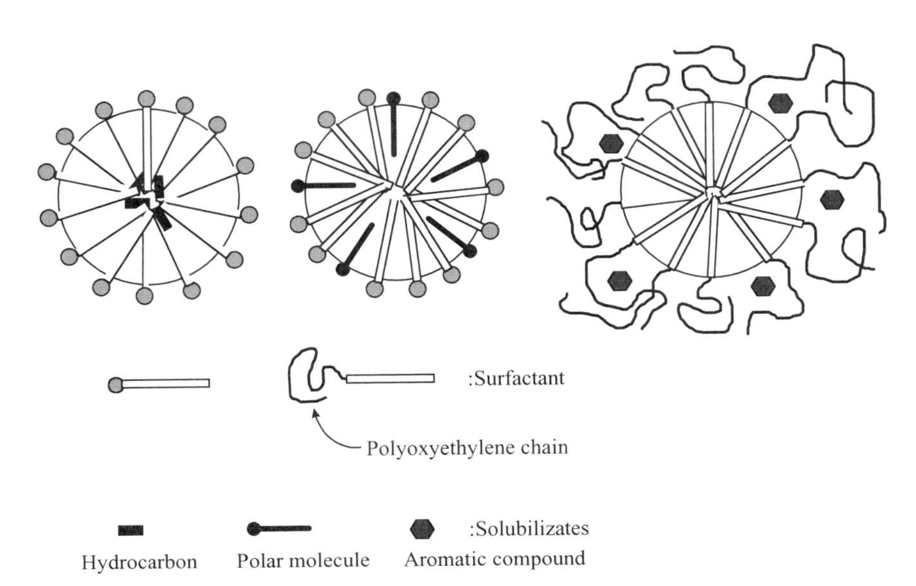

Figure 3.36 Schematic illustration of the solubilized locations of some organic compounds in the micelle. Nonpolar, polar, and aromatic compounds are solubilized in their most preferential positions in the micelle.

as well as in the hydrocarbon core since they are soluble in polar organic solvents.

The solubilization behavior of nonionic surfactants is quite unique in its temperature dependence. Figure 3.37 shows typical examples of such behaviors. The solubilized amount of some hydrocarbons by 1 wt % polyoxyethylene nonylphenyl ether is plotted as a function of temperature. The solubilization capacity of the nonionic surfactant increases dramatically in the narrow temperature range close to its cloud point. Solubilized amount of water into the oil phase containing 1 wt % nonionic surfactant of the above also sharply increases near the cloud points. These figures were already mentioned in Section 3.2.3.*d*, Figure 3.13. One liquid phase denoted I_w in Figure 3.13 corresponds to Figure 3.37, and the I_o region is the oil phase solubilizing water. The solubilization of oil (water) into water

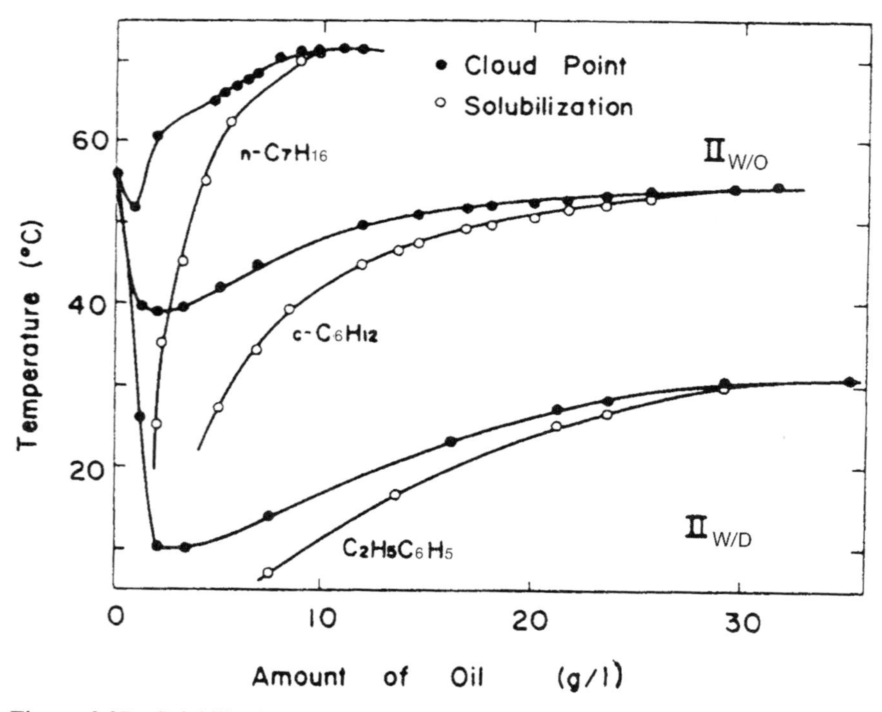

Figure 3.37 Solubilization amount of hydrocarbons into 1 wt % aqueous solution of poly-oxyethylene (p = 9.2) nonylphenyl ether as a function of temperature (Shinoda and Becher, 1978b). Reprinted from "Principles of Solution and Solubility," p. 185 (1978) by courtesy of Marcel Dekker, Inc.

(oil) phase becomes markedly large just before (after) the emulsion type changes from O/W to W/O. At lower temperatures the hydrophilicity of a nonionic surfactant is large, and the aggregate structure tends to get convex toward the hydrophilic group and to form O/W emulsions; vice versa at higher temperatures. Phase inversion temperature (PIT) is just the boundary between O/W and W/O emulsion, and the surfactant aggregates prefer neither convex nor concave structure; i.e., they have a flat layer structure at PIT. As a consequence, the micelles become the largest at PIT, as does the solubilization capacity of the surfactant.

As will be described in Section 3.3.4.*b*, vesicles and liposomes incorporate some drugs and are used as carriers in drug delivery systems (DDSs). This incorporation of drugs by vesicles and liposomes is also regarded as a solubilization phenomenon.

d. Microemulsions

Microemulsion is misnamed: though called an "emulsion," it is actually a solubilization system. It is a thermodynamically stable state and is not phase-separated forever. The term *microemulsion* came from its historical background: It was discovered when an ordinary turbid emulsion transformed to a transparent solution on addition of a cosurfactant (such as medium- or long-chain alcohols) (Hoar and Schulman, 1943; Schulman *et al.*, 1951; Schulman *et al.*, 1959; Stoeckenius *et al.*, 1960). It was thought that the particle size of emulsion became very small (10–100 nm) due to the added alcohols, causing the emulsion to become transparent. Since people believed that this system was an emulsion having a very small droplet size, the system was named microemulsion. However, extensive research done later has made it clear that Schulman's microemulsion is actually a kind of solubilization system containing a large amount of oils (Gillberg *et al.*, 1970; Ekwall *et al.*, 1970; Shinoda and Kunieda, 1973; Shinoda and Friberg, 1975; Shinoda and Lindman, 1987).

Figure 3.38 shows the phase diagram of a 20 wt % aqueous sodium dodecyl sulfate/benzene/pentanol system (Ahmad *et al.*, 1974). One can see that the homogeneous L_1 phase has a very sharp peninsula-shaped region toward the benzene corner and that the solubilization amount of benzene is enhanced dramatically with the addition of pentanol. This narrow region of L_1 phase is just the microemulsion itself, formed by the added pentanol. Pentanol adjusts the HLB of the surfactant and enhances the solubilization of benzene. This behavior of solubilization is, therefore, essentially the same as that in the nonionic surfactant system shown in Figure 3.37. The difference is only that the HLB of the nonionic surfactant is changed by temperature, whereas that of the ionic one is changed by

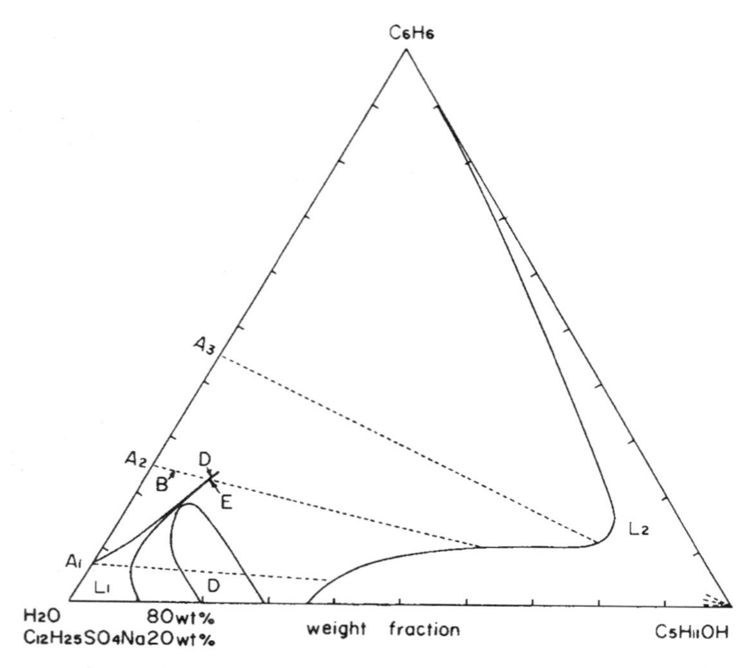

Figure 3.38 Phase diagram of the 20 wt % aqueous solution of sodium dodecyl sulfate/benzene/pentanol system at 30°C (Ahmad *et al.*, 1974).

the addition of a cosurfactant (pentanol). So we can recognize that the sharply elongated solubilization limit of hydrocarbons in Figure 3.37 can be also regarded as a microemulsion.

There are three types of microemulsion: O/W (oil solubilized in water), W/O (water solubilized in oil), and bicontinuous. The bicontinuous phase of microemulsions appears between the O/W and W/O types. We can transform these structures in the same system by changing the conditions. Fig. 3.39 shows the results of an NMR self-diffusion study, indicating the continuous transformation from O/W to W/O through a bicontinuous structure by changing the salinity of the solutions (Guering and Lindman, 1985). In the O/W microemulsions, water in the continuous phase can migrate far away, but the oil, which is entrapped in the very limited space of the solubilized droplets, cannot migrate. In the W/O type microemulsions, the situations are, of course, reversed.

The self-diffusion constants of water and toluene in Figure 3.39 reflect these structural features. The data of surfactant (SDS) diffusion are quite interesting. The surfactant molecules adsorb at oil/water interfaces and can migrate rapidly

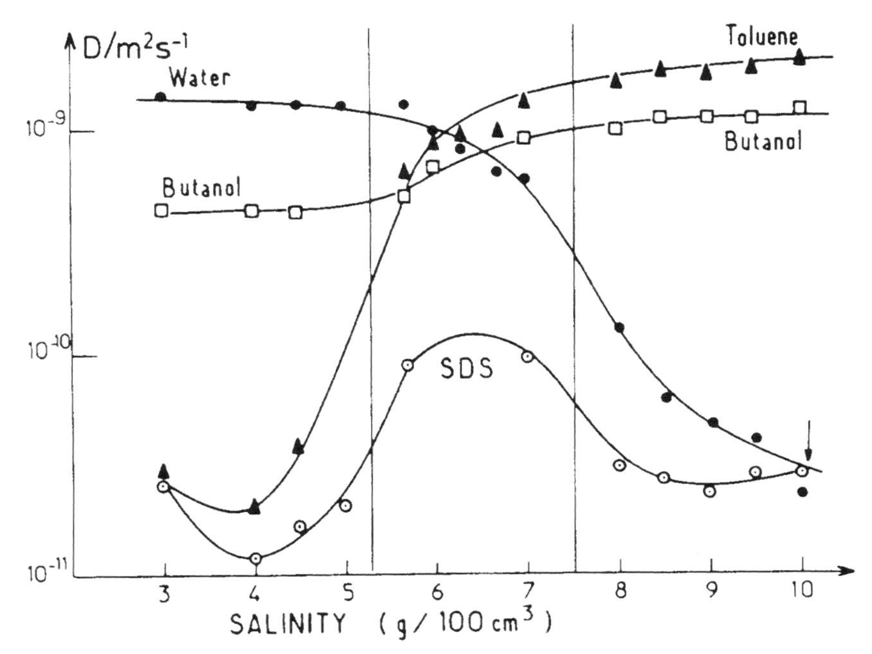

Figure 3.39 Self-diffusion coefficients of each component in the microemulsion system of sodium dodecyl sulfate (SDS)/butanol/toluene/brine (Guering and Lindman, 1985). Salinity of the brine was varied between 3–10 wt %. Reprinted with permission from *Langmuir,* **1,** 464 (1985), American Chemical Society.

only in the bicontinuous phase. In both the O/W and W/O structures, the molecules are fixed at the interfaces of tiny solubilized droplets and have small self-diffusion coefficients.

For more information on microemulsions, see, e.g., Prince, 1977; Robb, 1982; Rosano and Clausse, 1987; Friberg and Bothorel, 1987; Bourrel and Schechter, 1988; and Chen and Rajagopalan, 1990.

3.3.3 LIQUID CRYSTAL FORMATION

a. Liquid Crystal Formation in Surfactant Systems

A surfactant is one of the most typical substances that form lyotropic liquid crystals. Liquid crystals of surfactants are constructed of micelles that are periodically arranged in long-range order in the whole space of the solution. The structure of the liquid crystals is governed mainly by the structure of the micelle itself. Globular, cylindrical, and platelike structures of micelles result in cubic, hexagonal, and lamellar structures of liquid crystalline phases of surfactant. Figure 3.40 shows the phase diagram of a potassium myristate/water system (Ekwall, 1975). In the isotropic solution region, micelles are randomly dispersed and the solution is optically isotropic. A liquid crystal phase called the *middle phase* starts to appear at about 25 wt % of the surfactant, and another called the *neat phase* is formed at about 60 wt %. The structure of these liquid crystals are illustrated in Figure 3.41 together with a third phase, the *reversed middle phase.* Long cylin-

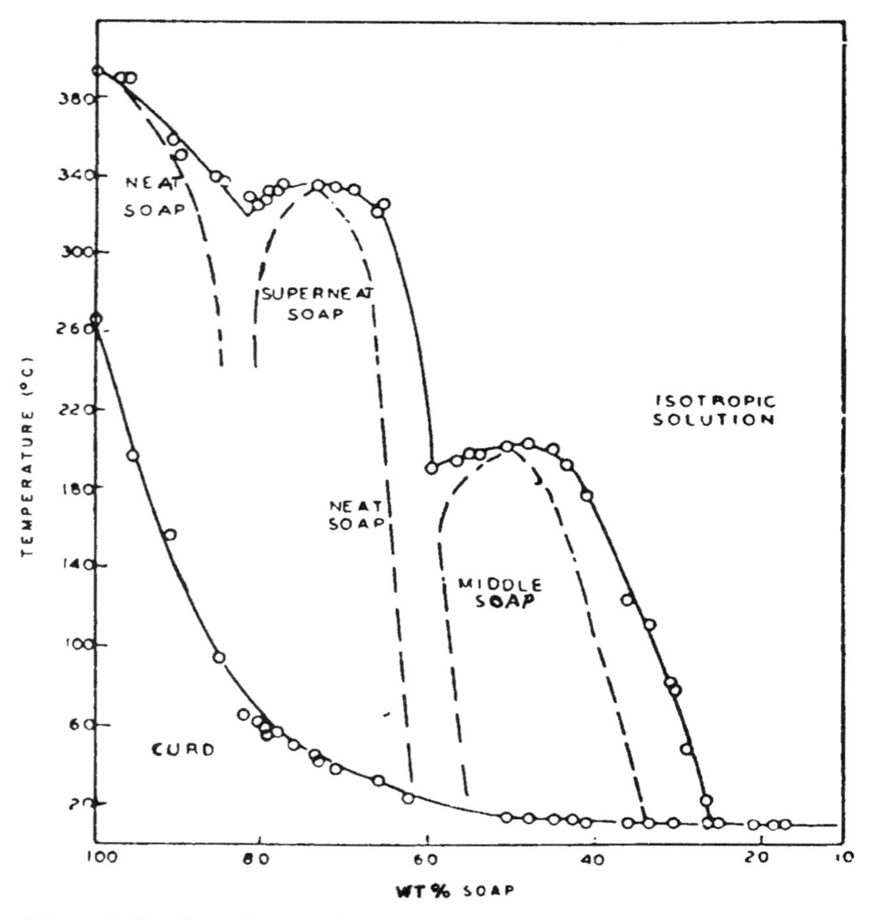

Figure 3.40 Phase diagram of a potassium myristate/water system (Ekwall, 1975).

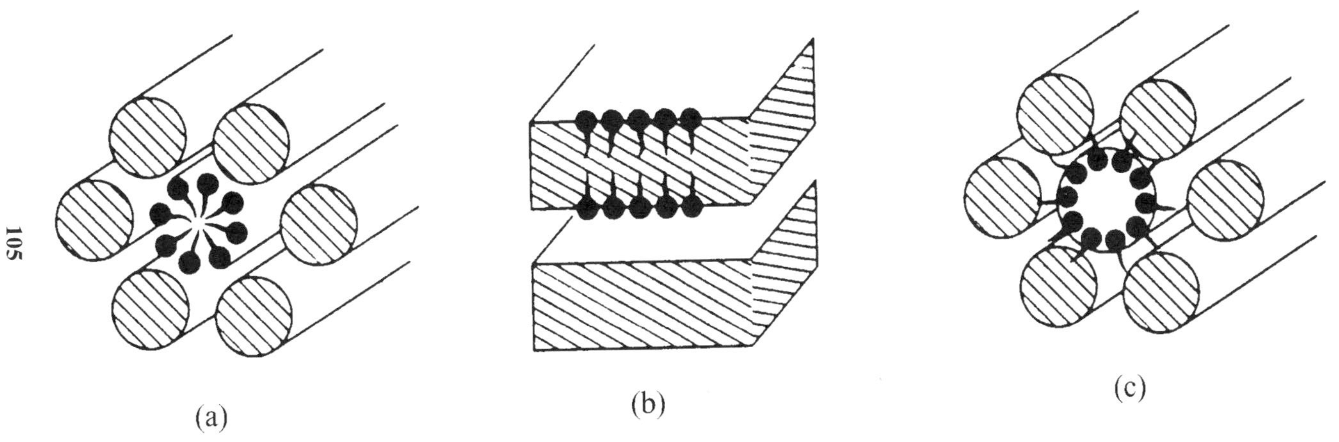

Figure 3.41 Schematic illustration of the structures of typical liquid crystals of surfactant in water: hexagonal (middle) phase (a), lamellar (neat) phase (b), and reversed hexagonal (middle) phase (c). Reprinted with permission of The Chemical Society of Japan.

drical micelles are packed in hexagonal symmetry in the middle phase. In the neat phase, platelike micelles are stacked periodically, forming a sandwich structure together with water layers. This sandwich structure is called the *lamellar structure,* or simply *lamella.* The reversed hexagonal phase is obtained in special surfactants with small hydrophilic head groups and are constructed of reversed rodlike micelles in which surfactant molecules are arranged orienting their hydrophobic tails outside. These three liquid crystal phases exhibit optical anisotropy and give beautiful and characteristic microscopic textures under crossed polarizers, which are often used to identify the type of liquid crystal.

Some surfactants provide completely different phase diagrams from Figure 3.40. Two examples are shown in Figure 3.42(a) (Small, 1968) and 3.42(b) (Kunieda and Shinoda, 1978). Both surfactants—egg yolk lecithin and ditetradecyldimethylammonium chloride—are double-chain ones. In these two phase diagrams, there is no isotropic and hexagonal (middle) region; only a lamellar phase exists. As mentioned in Section 3.3.2.*b*, double-chain surfactants tend to form platelike micelles due to the packing conditions. Then the lamellar structure is maintained even when the liquid crystal is diluted. The lamellar phase separated from and dispersed in water is called *multilamellar* (or multilayer) *vesicles* which will be discussed later in detail. They are basically spherical in shape and have an onion-like structure with repeatedly stacked bilayer membranes.

There exists a series of liquid crystalline phase of cubic symmetry that does not show any optical anisotropy (Lindblom and Rilfors, 1989, 1992). Two basic types of cubic phase are known: one is discrete and the other is bicontinuous. Figure 3.43 shows typical examples of the two structures. The discrete cubic phase consists of discrete micelles arranged in long-range order. In the bicontinuous phase, bilayer membranes infinitely expand and separate the water phase into two parts; both the water and surfactant phases are continuous in the whole solution.

Several types of bicontinuous structure are known, but in every case the bilayer membranes are of infinite periodic minimal surfaces. A *minimal surface* is defined as one that has zero mean curvature at any point. Readers can recognize this situation in Figure 3.43(a). Every point on the bilayer membrane is in something like a saddle point at which two principal curvatures of positive and negative are cancelled out. If some surface is curved, the pressure between both sides separated by the surface must be different (see Section 3.2.1.*c*). The pressure between the outside and the inside of the bilayer membrane must be identical, and the mean curvature of the membrane must be zero at any point. The most simple infinite periodic minimal surface is a flat lamella. The governing factor for the formation of the lamellar or the bicontinuous cubic phase from the same bilayer membrane of a minimal surface has not been elucidated.

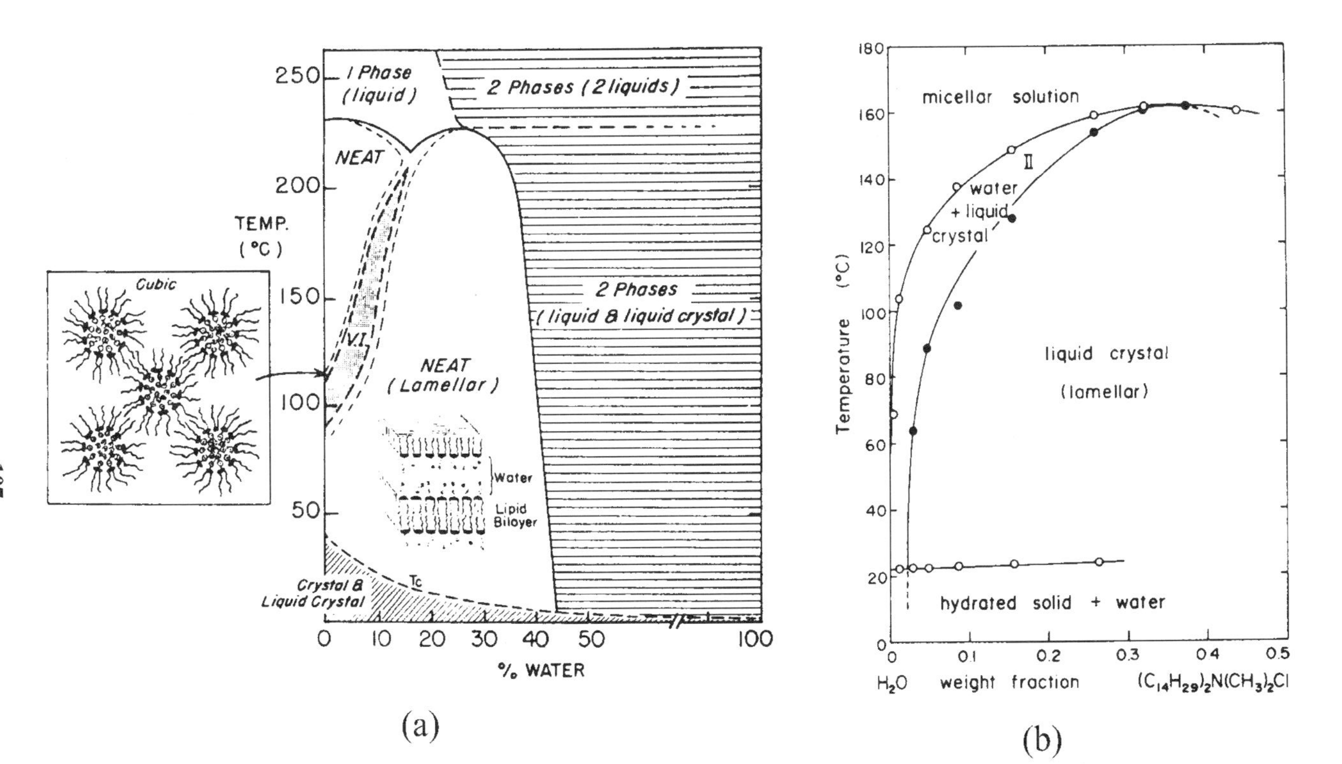

Figure 3.42 Phase diagrams of egg yolk lecithin/water (a) (by permission of D. M. Small, *J. Am. Oil Chem. Soc.*, **45,** 108 (1968)) and ditetradecyldimethylammonium chloride/water (b) (Kunieda and Shinoda, 1978, reprinted with permission from *J. Phys. Chem.*, **82,** 1710 (1978), American Chemical Society) systems. Note that the scale of the abscissa is opposite in (a) and (b).

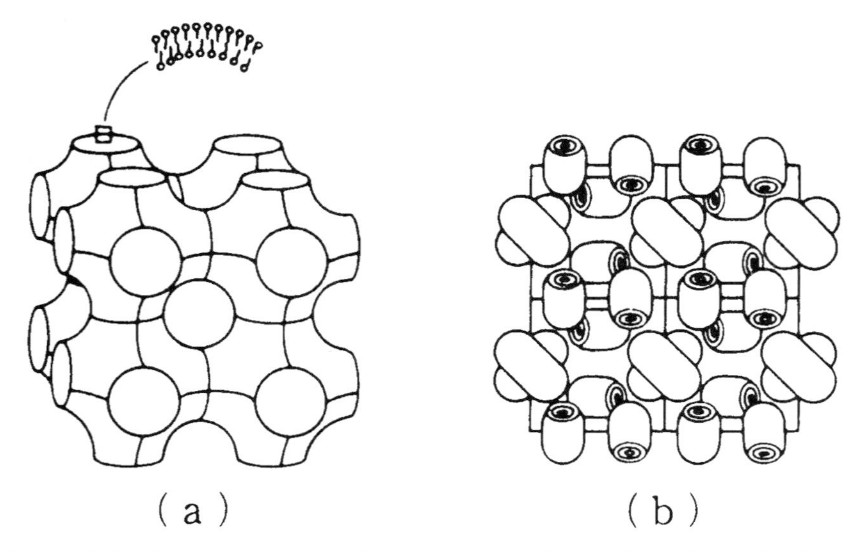

Figure 3.43 Schematic representation of some cubic phases of a surfactant liquid crystal: bicontinuous (a) and discrete (b). Reprinted with permission of The Chemical Society of Japan.

b. Interesting Phenomena in Surfactant Liquid Crystalline Phases

As pointed out in the previous section, surfactant solutions show various kinds of liquid crystalline structure. In this section, we discuss some interesting phenomena exhibited by the unique structures of surfactant liquid crystals.

The iridescent coloration phenomenon is known in dilute (1–2 wt %) aqueous solutions of some surfactants and surfactant mixtures, notwithstanding that such agents are not dyes and do not have any chromophores in their molecules (Tsujii and Satoh, 1992 as a review). Characteristic features of this iridescent solution are as follows.

1. The iridescent color appears only in the narrow concentration range of about 1–2 wt % irrespective of the type of surfactant.
2. The color shifts to the blue side with increasing concentration of the surfactant.
3. The color changes by the angle between the directions of light irradiation and an observer.

4. The color disappears by mechanical agitation as well as addition of inorganic salts.
5. The color shows beautiful luster.

Properties (2)–(4) strongly indicate that the color results from the higher-order structure of surfactant. The color change, especially by observing direction, suggests the diffraction phenomenon as the origin of the color appearance.

What kind of higher-order structure gives this diffraction phenomenon of visible light? In aqueous iridescent solutions of alkenylsuccinic acid, the higher-order structure was elucidated mainly by small-angle X-ray and UV diffraction techniques (Satoh and Tsujii, 1987). A first-order diffraction pattern of UV light was clearly observed, and the spacing distance calculated from the diffraction pattern gave a reasonable explanation of the iridescent color. But unfortunately, the actual structure of the solutions could not be determined because of the lack of higher-order patterns of UV diffraction. Small-angle X-ray diffraction experiments were, then, made for a higher concentration range (10–60 wt %) of the surfactant. Clear diffraction patterns up to the 4th order were observed in this case, and one series of Bragg reflections corresponded to the spacing with an interrelation of 1:1/2:1/3:1/4. These results indicate that the structure of alkenylsuccinic acid is lamellar at high concentrations.

Thus, we know the structure of alkenylsuccinic acid solutions of higher concentrations. Is the lamellar structure also maintained in dilute iridescent solutions? If the lamellar structure is formed uniformly in the entire space of the solutions, the interplanar spacing, d, can be related to the weight fraction of the surfactant, C, as (Luzzati *et al.*, 1960)

$$d = \left[\frac{(1-C)}{C} \frac{\rho_1}{\rho_2} + 1 \right] d_1 \tag{3.20}$$

where ρ_1 and ρ_2 are the densities of the surfactant and water layers, respectively, and d_1 is the thickness of the surfactant layer. The plot of the interplanar distance, d, against $(1 - C)/C$ is shown in Figure 3.44. The plots obtained from both UV diffraction of dilute solutions and X-ray diffraction of concentrated solutions are linearly related to $(1 - C)/C$ and are even identical. The thickness of the surfactant layer can be estimated from the y-intercept of this plot to be 4.1 nm. These results clearly indicate that the iridescent solutions also consist of a lamellar liquid crystalline phase constructed of single bilayer membranes. The iridescent color appears by the interference of reflected light due to the periodic structure of the single bilayer membranes, i.e., the diffraction of visible light. Whereas the

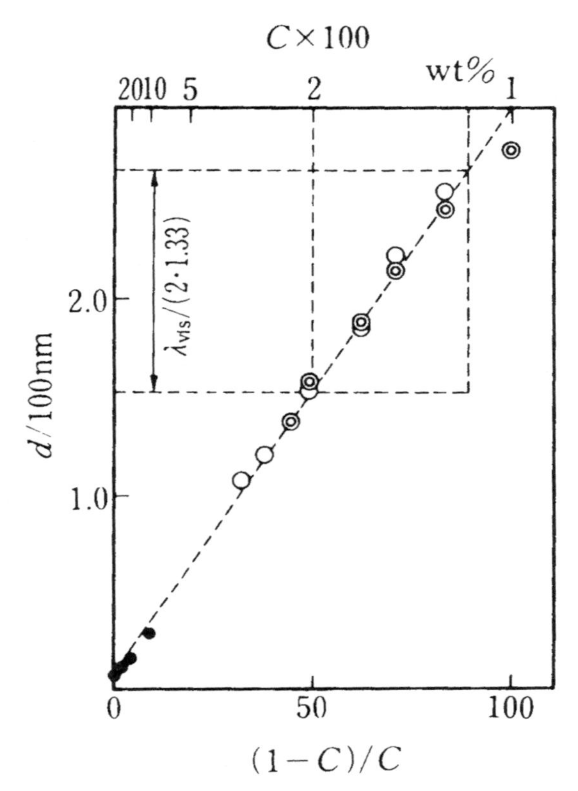

Figure 3.44 The plot of interplanar distance, *d*, against (1 − *C*)/*C* for the iridescent lamellar liquid crystal of alkenylsuccinic acid (Satoh and Tsujii, 1987). Reprinted with permission from *J. Phys. Chem.*, **91,** 6629 (1987), American Chemical Society.

lamellar liquid crystals of ordinary water-soluble surfactants transform to hexagonal liquid crystals and globular micelles when diluted, surfactants such as alkenylsuccinic acid maintain their lamellar structure until very low concentrations (1–2 wt %) and show this interesting iridescent phenomenon.

The iridescent color of the lamellar liquid crystalline phase is changed only by the surfactant concentrations, as mentioned above. There is another iridescent system in which the color can be changed by temperature or ethanol concentration as well as surfactant concentration. This system is a ternary mixture of a double-chain surfactant (triethanolammonium dihexadecyl phosphate:DHP)/water/ethanol (Yamamoto *et al.*, 1996), which shows the iridescent color only in the gel phase of the surfactant below its Krafft point. The color change with temperature can be interpreted by LCST phase separation between the ordered (lamellar) and disordered bilayer phase (Figure 3.23), which is very interesting as a fundamental problem on the interactions of colloidal particles mentioned previously (see Section 3.2.4.*e*). The diagram of the LCST phase separation is shown in Figure 3.45. This phase separation has never been observed in the iridescent lamellar liquid crystalline phase. It is of profound interest why the phase separation occurs only in the system of solid bilayer dispersions (gel phase) and not in the liquid bilayer systems (liquid crystalline phase).

Iridescent periodic lamellar structures of a polymerizable surfactant have been immobilized in polymer hydrogels (Naitoh *et al.*, 1991). These novel hybrid gels show some unique properties that cannot be obtained by individual polymer gels or bilayer membranes. The iridescent color can be changed in this case by controlling the swelling degree of the gels (Hayakawa *et al.*, 1997; Tsujii *et al.*, 1997a). The hybrid gels are much improved in mechanical properties and exhibit

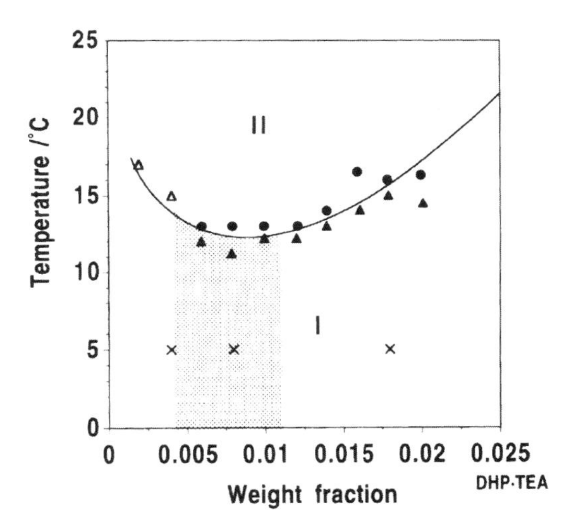

Figure 3.45 Phase diagram of the DHP gel phase as functions of temperature and surfactant concentration in an ethanol/water mixture (24/76 in weight) (Yamamoto *et al.*, 1996). Gradual phase transition takes place from the ordered lamellar to the disordered bilayer state in the shaded region. Freeze-fracture TEM was observed for the samples obtained at the × marks; the two corresponding to the highest and the lowest concentrations were shown in Figure 3.23. Reprinted with permission from *Langmuir,* **12,** 3134 (1996), American Chemical Society.

a macroscopic anisotropy when polymerized under shear flow (Tsujii *et al.*, 1997a).

One more interesting phenomenon is the "ringing gel," which is given by a cubic liquid crystalline phase (Oetter and Hoffmann, 1989). This transparent jellylike liquid crystal is obtained in a narrow region of the ternary phase diagram of an amphoteric surfactant (tetradecyldimethylamine oxide)/water/hydrocarbon. When its flask is hit with the palm of a hand, this liquid crystal emits in low tone; hence the name "ringing gel." The sound frequencies are in the range of 1000 Hz, and the damping time constant is about 1 sec. The structure of this gel-like liquid crystal was analyzed by small-angle neutron and X-ray scattering techniques and was determined to be a primitive cubic phase consisting of globular aggregates. The relatively long damping time of the sound is presumably because of the energetic elasticity originated from the energetically stabilized cubic lattice. This origin of the elasticity is completely different from the entropic one in solutions of long cylindrical micelles (see Section 3.3.2.*b*).

3.3.4 BILAYER MEMBRANES

a. Fundamental Properties of Bilayer Membranes as a Surfactant Micelle

A plate-shaped micelle is a membrane with a thickness of 4–5 nm consisting of two molecular layers, frequently called a *bilayer membrane.* Figure 3.46 illus-

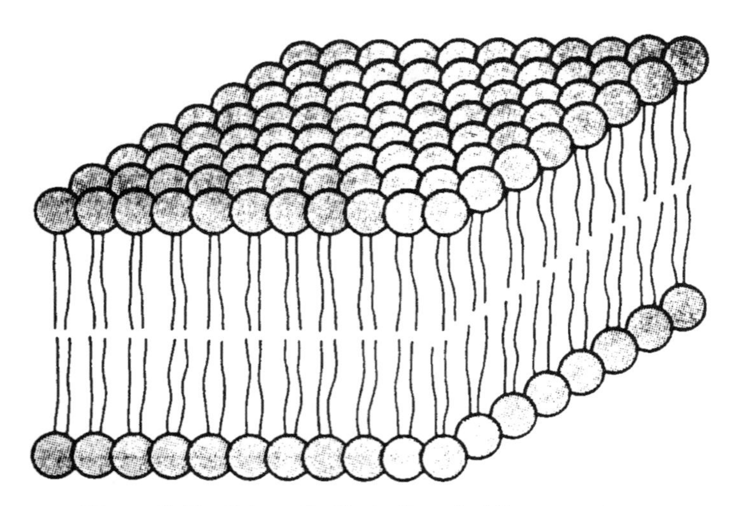

Figure 3.46 Schematic illustration of a bilayer membrane.

trates schematically such a membrane. In bilayer membranes, surfactant molecules are arranged in two contact layers orientating their hydrophilic head groups toward the water phase. The bilayer membranes formed with double-chain surfactants are stable (not transformed to rodlike and/or globular micelles) even in their diluted solutions; these attract special attention as a model of biomembranes. In this section, we start to discuss the fundamental properties of bilayer membranes from the viewpoint of surfactant micellar solutions.

Critical micellization concentration (CMC) and Krafft point are two fundamental physical quantities characterizing the surfactant (as mentioned previously). We will discuss the formation of bilayer membranes from these viewpoints. Does a "critical bilayer-membrane-forming concentration" exist? From a purely theoretical standpoint it should be present, but it is practically zero. Let us roughly estimate this concentration. It is known that the CMC values of a homologous series of surfactant decrease by one-third with the addition of one methylene group in the surfactant molecule. This phenomenon has been called *Traube's rule* (Traube, 1891; Nakagawa and Shinoda, 1963). According to Traube's rule, a double-chain surfactant should have a smaller CMC by $(1/3)^{10-20} \approx 10^{-5}$–$10^{-10}$ than a normal single-chain agent since one more hydrocarbon chain (10–20 methylene units) is attached to its molecule. As the CMC of a normal surfactant is about 10^{-3} mol/l, the CMC values of the double-chain ones are in the order of 10^{-8}–10^{-13} mol/l (i.e., practically zero). It is quite important for the double-chain surfactants and lipids that their CMC values are so small. It means that the bilayer membrane structure is not broken even when the surfactant is diluted, so the membrane works in the variety of environmental conditions of biological systems and practical applications.

The gel–liquid crystalline phase transition temperature, T_C, is a Krafft point (see Section 3.3.1.*b*). In an ordinary surfactant solution, a (hydrated) crystalline solid of surfactant transforms into small globular micelles at the Krafft point to form a transparent solution. In the case of bilayer-forming surfactant, on the other hand, the micellar phase is a (lamellar) liquid crystal made of platelike micelles having an infinite aggregation number, and the solid phase is a gel or a coagel (real crystal) phase. The gel phase is translucent and does not flow even if the vessel is turned upside down. It is often in a thermodynamically metastable state and can be transformed into a coagel phase by standing the sample for long time or annealing it carefully. The gel phase is a two-dimensional crystal, as mentioned in Section 3.3.1.*b*. The surfactant molecules start to move at the T_C in their molecular position and hydrocarbon chain. This discontinuous change of molecular motion results in the sudden alteration of membrane properties. Permeation of some substances through the bilayer membrane is an example of such a change. It is important for practical applications of bilayer membranes to utilize the difference in membrane properties above and below the gel–liquid crystalline phase transition temperature T_C.

b. Vesicles and Liposomes

Bilayer membranes form closed globular structures in water, as illustrated in Figure 3.47. These hollow bodies of surfactants and lipids are called *vesicles* or *liposomes*. There are two kinds of vesicles and liposomes: one is multilamellar (or multicompartment or multilayer) and the other is unilamellar (or single-compartment or single-bilayer). The multilamellar vesicles are constructed of repeatedly stacked bilayer membranes forming an onion-like structure and have a diameter of the order of μm. As a consequence, a dispersion of multilamellar vesicles is turbid. The multilamellar vesicle is exactly the same as the lamellar liquid crystal separated from and dispersed in aqueous phase (see Section 3.3.3.*a*). The unilamellar vesicle, on the other hand, consists of a single-bilayer membrane and usually gives a transparent solution.

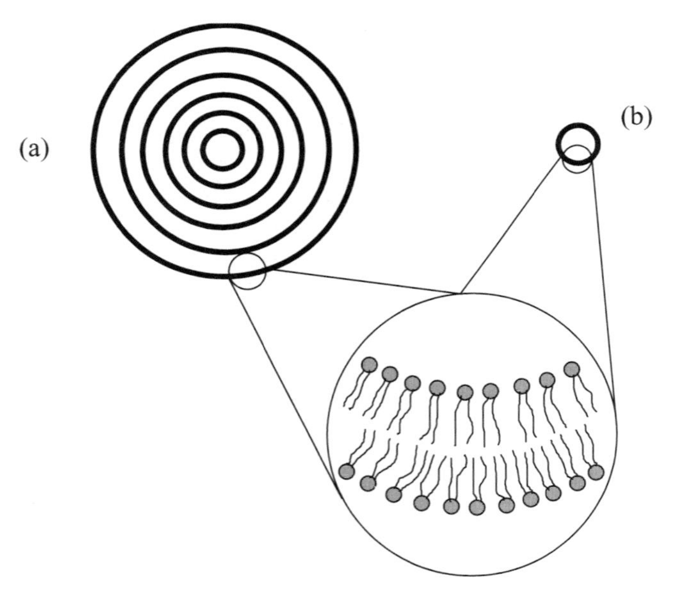

Figure 3.47 Schematic illustration of liposomes (vesicles): multilamellar (a) and unilamellar (b).

Multilamellar vesicles are spontaneously formed when some lipids (such as a lecithin) are put into water. Unilamellar vesicles are commonly prepared by ultrasonication of the multilamellar vesicles (Bangham *et al.*, 1965; Bangham *et al.*, 1974). The unilamellar vesicle can be also obtained by other methods. The vesicle-forming lipids dissolved in a water-miscible organic solvent are injected into water to form vesicles in an aqueous medium (Batzri and Korn, 1973; Kremer *et al.*, 1977). An easily volatile solvent (such as ether) is also utilized for the injection method of vesicle formation (Deamer and Bangham, 1976; Schieren *et al.*, 1978). The mixed micelles of the lipids and a water-soluble hydrophilic surfactant transform into vesicles when dialyzed toward water (Zumbuehl and Weder, 1981; Ueno *et al.*, 1984). The unilamellar vesicles obtained by means of ultrasonication have a size of several tens of nm, but much larger ones (more than μm in size) can be formed by the other methods. The vesicle of larger size is sometimes called a *large unilamellar vesicle* (LUV) and the smaller a *small unilamellar vesicle* (SUV) when necessary.

When the liposome was first found (Bangham *et al.*, 1965), it was thought that only natural lipids of the important component of biomembranes could form such a special structure. More than ten years later, however, a totally synthetic surfactant—dioctadecyldimethylammonium chloride—was shown to form the same vesicle structure as that of natural lipids (Kunitake and Okahata, 1977; Deguchi and Mino, 1978; Tran *et al.*, 1978). Liposomes, then, are regarded as one of the aggregated structures of surfactant formed by pure physicochemical means. The term *vesicle* is more general and broader than *liposome*: The vesicle formed with natural lipids is a liposome.

The unilamellar vesicle prepared by ultrasonication is not thermodynamically stable; it is sooner or later transformed to a multilamellar vesicle (Franses *et al.*, 1982; Nozaki *et al.*, 1982). However, when surfactant systems show spontaneous vesicle formation, these unilamellar vesicles are thermodynamically stable (Gabriel and Roberts, 1984; 1986; Talmon *et al.*, 1983; Ninham *et al.*, 1983; Miller *et al.*, 1987).

As previously pointed out, the multilamellar vesicle is exactly the same thing as the lamellar liquid crystalline phase separated out of and dispersed in an aqueous phase. There is an opposite type of separated lamellar liquid crystal that is dispersed in an oil phase. This lamellar liquid crystalline phase consists of a surfactant, a sugar ester of fatty acid, and an oil in the presence of a small amount of water, and it is dispersed in the oil phase. This dispersed liquid crystal in the oil phase can be called a *reverse vesicle,* just as the reverse micelle is defined (Kunieda *et al.*, 1991, 1993, 1994).

c. Bilayer Membranes as an Artificial Biomembrane

Bilayer membranes have attracted much attention from the beginning of their research because their structure is like that of biomembranes. Two approaches have been taken in their study: One is to model biomembranes and the other is to apply them as artificial biomembranes in some practical way. The latter will be discussed in the next chapter.

Bilayer membranes can be regarded as the simplest and the most essential structure of biomembranes. In real biomembranes, there are a number of materials other than lipids, and the essential function of the lipids is hardly seen. People utilize the lipid bilayer membranes to see the most simple functions of a biomembrane. For example, the permeation behavior of biologically important compounds through lipid bilayer membranes is actually similar to that through biomembranes (Bangham *et al.*, 1965, 1974; Klein *et al.*, 1971).

Transportation of materials through biomembranes is done by two mechanisms: One is direct passing through the lipid membrane and the other is a protein-assisted permeation mechanism. Bilayer membranes are, of course, a model of the first mechanism. Permeation experiments are performed as follows. Liposomes are formed in aqueous solutions containing a marker compound and then separated from the medium solution by, say, a gel-filtration method. The marker molecules inside the liposomes leak out and are detected by a proper technique. The permeability of biologically important substances such as various ions, amino acids, saccharides, etc., is measured this way. Selective permeation is of special interest in these permeability studies. Potassium ions permeate selectively and acceleratedly through a liposomal membrane when an antibiotic, valinomycin, is present (Blok *et al.*, 1974a, 1974b). This is a good example of liposomes as an artificial biomembrane.

Permeability control is also possible in the bilayer membranes of totally synthetic surfactants. The bilayer membranes of dioctadecyldimethylammonium chloride (DOAC) put in the pores of thin nylon capsules were tested (Okahata *et al.*, 1983). Figure 3.48(a) shows a schematic illustration of a nylon capsule coated with the DOAC bilayer membranes. Leakage of NaCl inside the capsule was detected by electric conductivity measurement and is plotted in Figure 3.48(b). One can see that the permeation of NaCl is fast at 45°C and very slow at 40°C. This temperature-switching of the permeability of NaCl is due to the gel–liquid crystal phase transition of DOAC bilayer membranes. The bilayer membranes are in a gel state below the phase transition temperature ($T_C \approx 42$°C) and form a barrier difficult for NaCl to pass through. At temperatures higher than T_C, on the

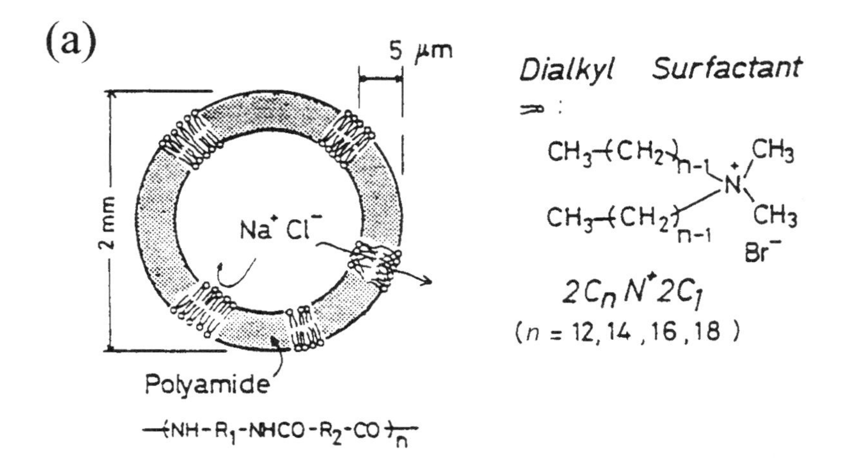

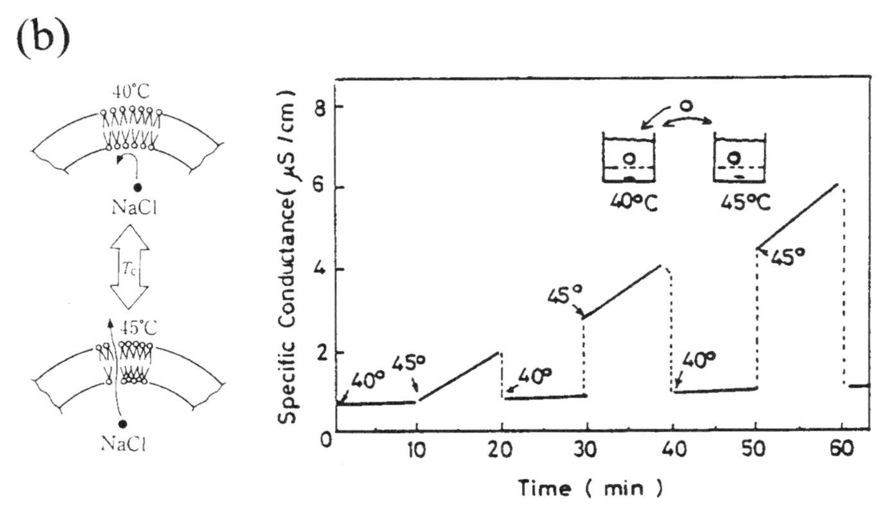

Figure 3.48 Schematic illustration of a nylon capsule coated with bilayer membranes of DOAC (a) and its permeation control by temperature (b) utilizing the gel–liquid crystalline phase transition of the surfactant (Okahata *et al.*, 1983). Reprinted with permission from *J. Am. Chem. Soc.*, **105,** 4855 (1983), American Chemical Society.

other hand, the NaCl can migrate into the membranes of liquid crystalline phase (Okahata *et al.*, 1983).

Recognition of cells and molecules is one of the most important functions of biomembranes. Two kinds of recognition mechanism are known at present: the interaction between saccharide chains and the antigen–antibody interaction. Glycolipids and glycoproteins present in the biomembranes play important roles in the saccharide chain interactions. The antigen–antibody interaction is, of course, an immunoreaction.

The liposomes bearing some saccharide chains actually show a recognition ability (Sunamoto, 1985). Polysaccharides (mannan, pullulan, amylopectin, dextran, etc.) hydrophobized with a cholesterol derivative coat a liposome surface anchoring the hydrophobic moiety inside the bilayer membranes. The polysaccharide chains situated at the surface of the liposome recognize and bind to a concanavalin A molecule, which is a receptor protein for saccharides. Concanavalin A has multiple binding sites for saccharide moiety, so the liposomes coagulate with each other through concanavalin A molecules. Coagulated liposomes coated with amylopectin are redispersed with the addition of mannose because mannose is a more favorite saccharide than amylose for concanavalin A. Such binding exchange does not occur when mannan is prebound to the concanavalin A, and no redispersion takes place. One can see from these results that the polysaccharide chains present at the surface of the liposomes work like an antenna for molecular recognition. The liposomes bearing an antibody for cancer cells are able to recognize the target cells and thus can be applied to a drug delivery system (DDS). This application will be discussed in the next chapter.

A number of membrane proteins are located in the biomembranes and show various functions that maintain the biological activities of the cell. Some of the proteins are just sitting on the surface of the membrane (peripheral proteins), some of them are buried in the membrane, and some are situated throughout the membrane (integral proteins). If these membrane proteins are transformed to liposomes as their native and active conformations, the reconstructed liposomal membrane can be regarded as an artificial biomembrane.

A clever method to achieve this has been developed recently. A special lipid called a *boundary lipid* is the key compound for this extraction method of membrane proteins. The membrane proteins are usually surrounded by the boundary lipid (such as sphingomyelin) and are stabilized in the biomembranes. They sometimes cannot perform their activities without the boundary lipid. The boundary lipid has amide bonds, instead of the ester bond of ordinary phospholipids, and may interact strongly with the proteins through hydrogen bonds. If we develop some artificial boundary lipid that can make the membrane proteins more stable, the liposomes containing the artificial lipid may be able to take the proteins from intact biomembranes. This idea has been realized by the artificial boundary lipid 1,2-dimyristoylamido-1,2-deoxy-phosphatidylcholine (DDPC) (Sunamoto *et al.*, 1990a, 1990b, 1992a, 1992b; Okumura *et al.*, 1994; Suzuku *et al.*, 1995). The DDPC molecule has two amide groups, as shown in Figure 3.49, and may form more hydrogen bonds with proteins than the natural boundary lipid.

The artificial boundary lipid (DDPC) was mixed with a normal phospholipid (e.g., 1,2-dimyristoylphosphatidylcholine; DMPC), and a mixed liposome was prepared. The mixed liposomes were coincubated with several kinds of intact cells and then separated again. The membrane proteins that transferred to the liposomes were analyzed. Figure 3.50 shows the effect of the DDPC content in the liposome on the extraction efficiency of proteins and the cell viability (a tumor cell; Balb RVD) (Sunamoto *et al.*, 1990b). The membrane proteins were extracted most effectively at the ratio of DDPC/DMPC = 80/20, and the cell viability decreased when many proteins were taken off. This indicates that the membrane proteins may be situated more stably in the liposomal membranes containing the artificial boundary lipid than in the intact cell membrane and thus transferred spontaneously to the liposomes. Furthermore, the liposomes bearing membrane proteins are expected to behave as an artificial cell. This extraction method will be a powerful tool in the future to study the membrane proteins and biomembranes themselves. It is interesting to note that membrane proteins also can be extracted from animal tissue. For example, taste receptor proteins of a bullfrog tongue were successfully transferred to DDPC/DMPC mixed liposomes, and the taste nerve response of the frog was remarkably depressed by this treatment (Nakamura *et al.*, 1994).

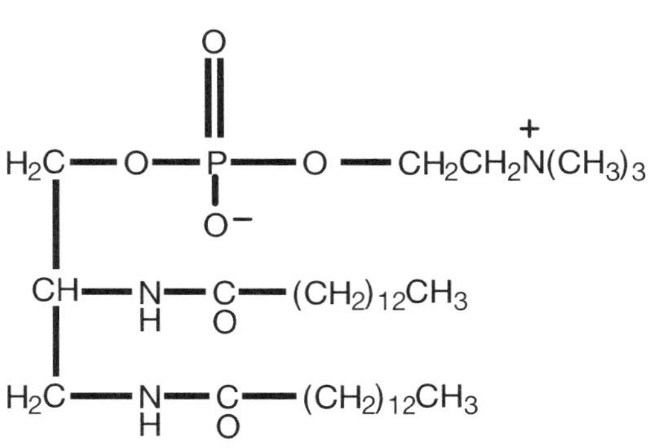

$$
\begin{array}{l}
H_2C-O-\overset{\displaystyle O}{\overset{\|}{\underset{\underset{\displaystyle O^-}{|}}{P}}}-O-CH_2CH_2\overset{+}{N}(CH_3)_3 \\
\;|\\
CH-\underset{H}{N}-\underset{O}{C}-(CH_2)_{12}CH_3 \\
\;|\\
H_2C-\underset{H}{N}-\underset{O}{C}-(CH_2)_{12}CH_3
\end{array}
$$

Figure 3.49 Molecular structure of an artificial boundary lipid, DDPC (Sunamoto *et al.*, 1990a).

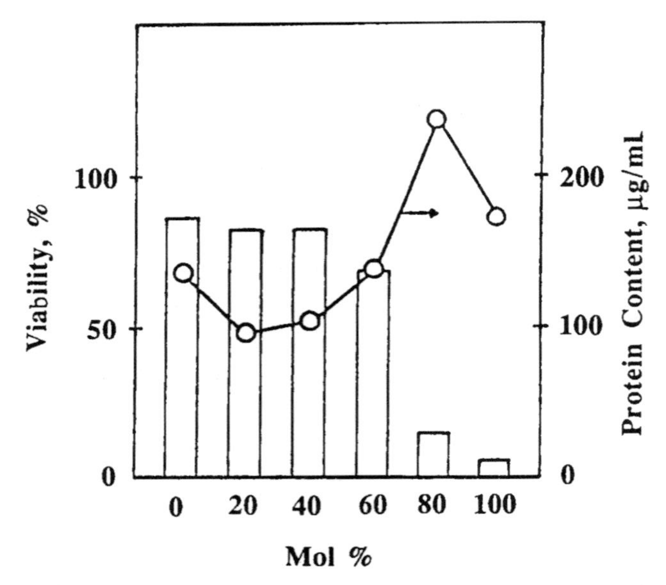

Figure 3.50 Effect of an artificial boundary lipid, DDPC, on the extraction of membrane proteins from a tumor cell, Balb RVD, to liposomes and on the cell viability (Sunamoto *et al.*, 1990b).

3.3.5 MONOLAYERS AND LB FILMS

a. Monolayers (A Two-Dimensional World of Materials)

One can sometimes see a beautiful iridescent film of car oil on a small water pool after the rain. This phenomenon is well understood as an interference of visible light reflected from the surface of the oil film and the interface between oil and water. To show the iridescence, the oil must be spread on the water surface until its thickness becomes submicrometer (the order of a wavelength of visible light). In this case, the wetting of the oil on water is complete and then must proceed until the monomolecular film of the oil on the surface area of the water pool is large enough. The monomolecular film, or monolayer, is thus obtained as a limit of wetting.

Benzene solutions of octadecanoic acid are frequently used in laboratories to obtain a monolayer of the acid. Interfacial tension between the solution and the water phase is low because the adsorption of octadecanoic acid at the interface orients its hydrophilic group toward the water and the benzene solution spreads on the water surface. The benzene evaporates after the solution spreads, and a monomolecular layer of the octadecanoic acid is left. This monomolecular membrane of water-insoluble amphiphilic compounds is often called a *Langmuir film* after the pioneer of this research field (Langmuir, 1916, 1917).

The monomolecular membrane remaining on the water surface behaves very interestingly. Figure 3.51 shows some examples of its so-called *π-A curve* (Adamson, 1982b). The abscissa, A, is the area occupied by one molecule of the amphiphile, and the ordinate, π, is the surface pressure, which is defined as the difference in surface tension between pure water and the monolayer surface. The surface tension of the monolayer surface is smaller than that of pure water, so the force of π per unit length acts toward the pure water side when two surfaces are in contact each other on the water phase.

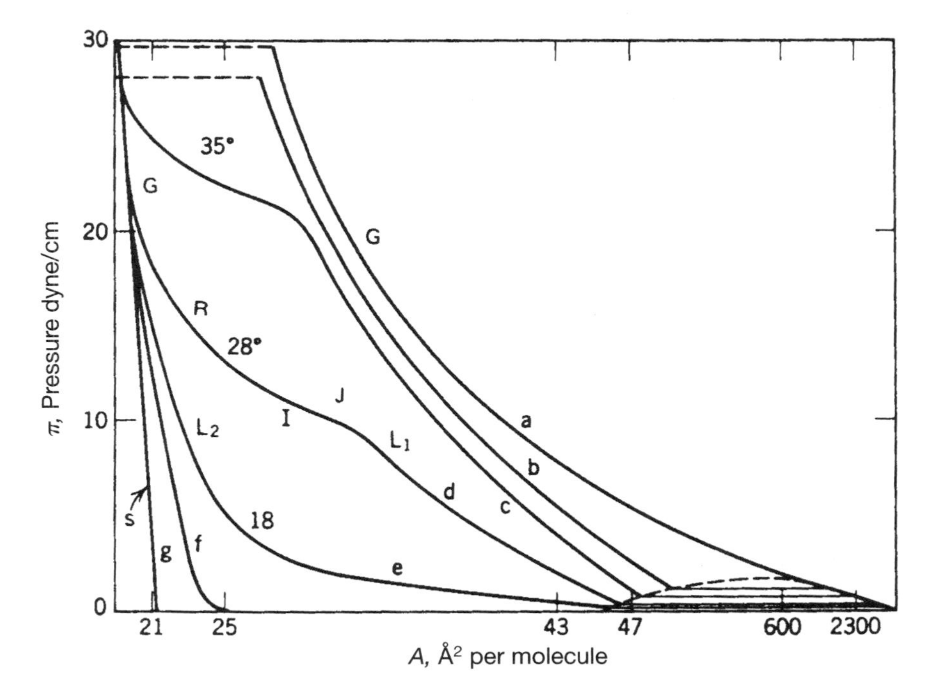

Figure 3.51 Surface pressure vs. area curves of a monomolecular layer spread on the water surface (Adamson, 1982b). These curves behave like the pressure–volume relationship in bulk material. "Physical Chemistry of Surfaces" 4th ed., Copyright © 1982. Reprinted by permission of John Wiley & Sons, Inc.

Surface pressure–surface area (π-A) curves are obtained by an apparatus such as is illustrated in Figure 3.52 (Moore, 1962). A shallow trough is used as a container for the pure water onto which the monolayer is spread. The water surface is separated into two parts by a movable barrier, and the monolayer is prepared on one side. The surface pressure π can be measured by the difference of surface tension between the two parts. The surface area is varied by moving the barrier, and the value of A is calculated. In recent years there have been many types of apparatus developed, but the basic principle is demonstrated in Figure 3.52.

Let us discuss the significance of the π-A curves shown in Figure 3.51. Curves a–g indicate that the surface pressure increases with decreasing surface area of the monolayer film, just like the pressure–volume relationship in three-dimensional bulk materials. One can see several different types of π-A curves in Figure 3.51. Let us take curve d as a typical one. Curve d coincides with curve a in the range of very large surface area at the lower right-hand side of the figure, then passes through a region where the surface pressure does not change even when the area decreases. The surface pressure starts to increase again beyond this region, The curve then exhibits two different slopes, indicating two kinds of state each of which has different compressibility. The curve finally becomes coincident with curve g, having very small compressibility.

The region in which the pressure does not change on decreasing the surface area reminds us of a phase transition phenomenon. In fact, amphiphilic compounds show two-dimensional transformation from a gaseous to a liquid state on the water surface. The liquid state transforms to a solid state when compressed more in the three-dimensional bulk phase. But in curve d, the liquid state L_1 transforms to another liquid state L_2 (range R–G in the curve), and then to the solid state s. The first liquid film (L_1) having greater compressibility, is called a *liquid expanded film* and the second one (L_2) is the *liquid condensed film*. The L_2 film changes to solid state, and curve d coincides with curve g. The surface area occupied by one amphiphilic molecule can be estimated to be 20–22 Å^2 from the extrapolation of curve g to the zero surface pressure. This value is close to that in the three-dimensional bulk crystal (18.5–20 Å^2). Sometimes only the L_1 film appears (curve b), and sometimes only the L_2 film appears (curve e).

There are two kinds of liquid state in the monomolecular films that are unlikely in the three-dimensional state; this is because the interaction between the amphiphilic molecule and water works in addition to that between the amphiphilic molecules themselves. In the liquid expanded films, the amphiphilic molecules essentially lie on the surface of water and stand up during compression of the film. In the liquid condensed film, on the other hand, the molecules are in almost their closest packing state standing perpendicularly to the water

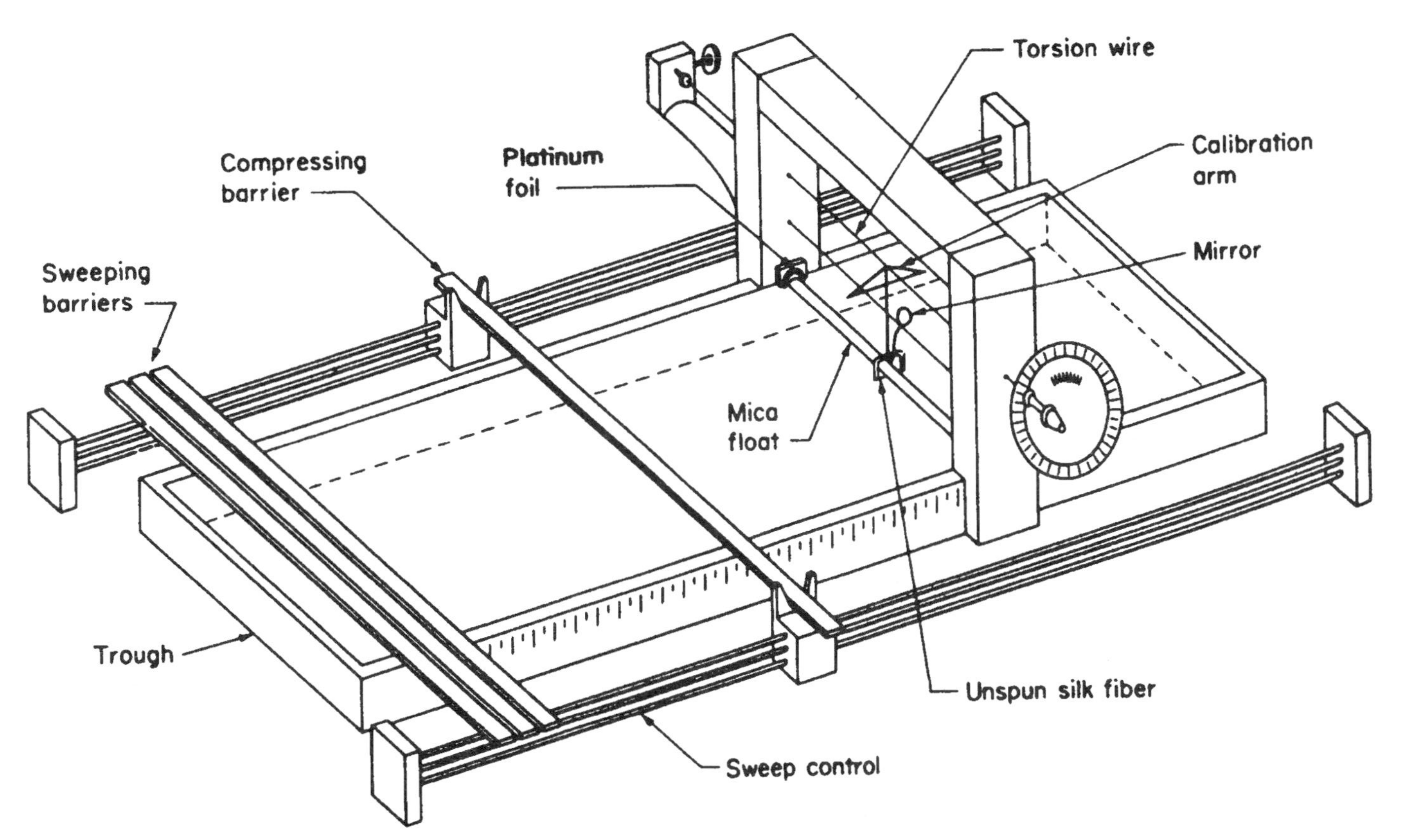

Figure 3.52 An apparatus to measure the surface pressure–surface area curves. Moore, Walter, "Physical Chemistry," © 1962. Reprinted by permission of Prentice-Hall, Inc.

surface and the rearrangement of the molecules—including the change of hydration structure—takes place when the film is compressed. It is readily understood from the structure difference between the two kinds of liquid film that the liquid expanded film shows much larger compressibility and molecule-occupied area than the condensed one. The area occupied by one molecule is 40–70 $Å^2$ in the expanded films and 22–30 $Å^2$ in the condensed ones. As mentioned above, the monolayers on the water surface exhibit two-dimensional phase changes similar to three-dimensional condensed matters. It is worth noting that the two-dimensional behaviors of materials can be more easily studied than three-dimensional ones, and so provide an ideal model for the adsorption of surface active substances.

It is not necessarily true that any kind of three-dimensional material shows all the phase transitions of gas → liquid → solid with increasing pressure. For example, helium never becomes liquid and solid at room temperature, and metals such as iron and copper never change to liquid and gas even when the pressure is reduced to a very low value. The situation is the same in the two-dimensional monolayers. In the case of amphiphilic compounds having weak attractive interaction between molecules, only the gaseous and liquid states can be observed; the solid state cannot. On the other hand, the solid state monolayer appears from the very low surface pressure when the molecular interaction is very strong.

This variety of phase changes is given in Figure 3.51. For strongly interacting amphiphiles, the solid state islands of the compound are formed on the water surface at low surface pressure. The surface pressure does not change even when the surface area is decreased, and only the distance between the islands becomes smaller. When the islands are in contact each other, the surface pressure starts to increase very steeply since the islands are a stiff solid; curves f and g in Figure 3.51 are examples. The monomolecular films so obtained by the above procedure are, of course, polycrystalline. To obtain a single crystalline solid film of large area, this preparation method is not adequate.

Only the substances soluble in organic solvents can be spread on the water surface as a monolayer. Another substrate is necessary to make a monolayer film of water-soluble materials such as proteins and nucleic acids. Mercury is one of the best substrates for this purpose (Yoshimura *et al.*, 1990; Kulkarni *et al.*, 1991). The much higher surface tension (~400 mN/m) of mercury compared to that of water (72 mN/m) is of great advantage in spreading the materials. Even aqueous solutions can be spread on it. But this advantage also means that the mercury surface is easily soiled, so special care is necessary for this technique.

b. LB (Built-up) Films (Molecular Architecture)

Monomolecular films (Langmuir films) on a water surface can be transferred to a solid surface. This process is schematically illustrated in Figure 3.53. A solid plate of glass, metal, mica, etc., is inserted into and/or pulled out of the water phase on which a monolayer of amphiphilic compound is formed at a constant surface pressure. The monolayer is then transferred onto the surface of the solid plate. Many sheets of films can be built up on the surface by repeating the process. The built-up film thus obtained is often called *Langmuir-Blodgett* (LB) *film* (after Miss Blodgett, a disciple of Langmuir and the inventor of this method).

There are three kinds of layered structure in the built-up films. The head-to-head (tail-to-tail) layered film, as shown in Figure 3.53, is the *Y film.* The head-to-tail structure, where the amphiphiles orient their hydrophobic groups toward the substrate solid surface, is the *X film,* and that with the hydrophobic groups toward the air side is the *Z film.* The initial structure of the film is governed by the preparation conditions, such as the hydrophobicity of the substrate and the nature of the amphiphilic compound, but is not necessarily a thermodynamically stable form. In such a case, the initially obtained layered structure is frequently observed to transform into a more stable one.

The technique to make LB films is quite sophisticated. The built-up films can be called *molecular architecture* as they are constructed with building blocks of each monomolecular membrane. It will require visionary technology in the next century to assemble the molecules artificially as we want, just like in this century we can assemble atoms to synthesize desired molecules (organic chemistry). The Langmuir-Blodgett method to build up the films seems to partly establish such a technology.

Let us show some interesting applications of this sophisticated technique. An insulator film of a one-molecular layer thickness may be useful in the field of electronics. In the dielectric of condensers, for instance, a thinner film can store more electric charges. LB films are also being studied for use as an insulator for Schottky junctions of MIM (metal/insulator/metal) and/or MIS (metal/insulator/semiconductor) devices, tunneling junctions, field effect transistors, and so on (Roberts, 1985, 1990). These studies are still just basic research and far from practical applications. An example of a more sophisticated application of LB films is the electrochemical photodiode (Fujihira *et al.*, 1985). A schematic illustration of the expected structure of the diode is shown in Figure 3.54(a). The parts denoted by A, D, and S in the figure are an electron acceptor, a donor, and a sensitizer, respectively. These are connected with an amphiphilic molecule in a chemi-

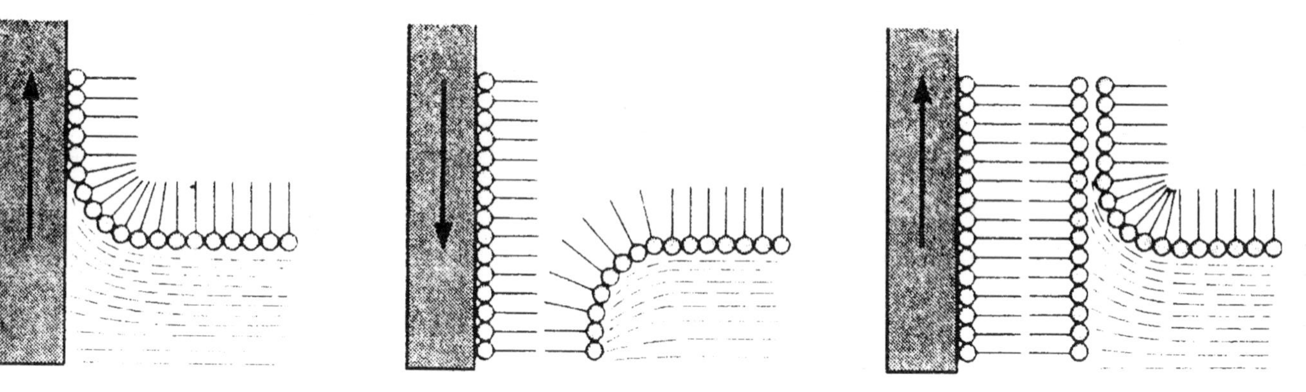

Figure 3.53 Schematic illustration of the LB film preparation. An example of Y-film is shown.

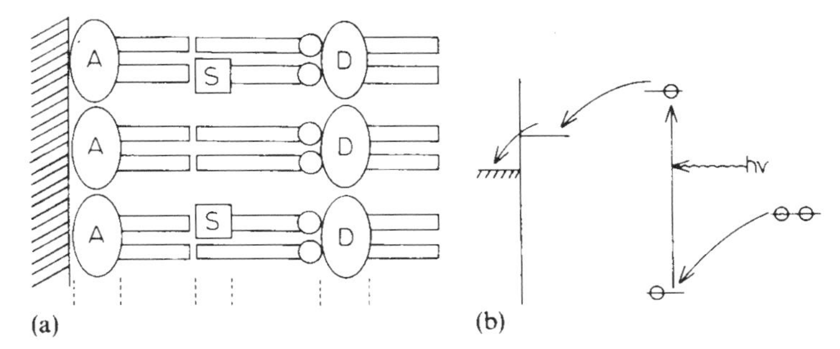

Figure 3.54 Schematic illustration of a molecular photodiode constructed by the LB technique (a), and its working mechanism (b). A, S, and D are the electron acceptor, sensitizer, and electron donor, respectively. Reprinted from *Thin Solid Films,* **132,** M. Fujihira *et al.*, p. 77 (1985), with permission from Elsevier Science.

cal bond and built up on a gold optically transparent electrode (Au-OTE) by the LB method as a well-organized molecular assembly. When light is irradiated to this film, the sensitizer molecule is excited and the electron is transferred to the acceptor molecule and then finally to the Au-OTE. The electron is supplied to the sensitizer from the donor molecule (Fig. 3.54(b)). Photoelectric current was actually detected using viologen, pyrene, and ferrocene moieties as A, S, and D, respectively.

The assembled architecture of molecules in the LB films also affects their chemical reactions. Configurational control in polymerization reactions is particularly interesting in relation to the theoretical molecular evolution in the primitive earth. Condensation polymerization of some amino acid esters of fatty alcohol has been accelerated in LB films, and it was found that the structure of the synthesized polypeptide depended on the arrangement of the monomer molecules (Fukuda *et al.*, 1983). Photopolymerization of 2,4-octadecadieneic acid in the LB films provided one more typical example of the configurational control resulting from the assembled structure of the monomer (Fukuda *et al.*, 1985). Two LB films of the monomer built up from the liquid expanded and the liquid condensed film gave the different structure in the polymer molecule. Polymerization reactions in a self-assembled structure of monomer molecules is also known in the bilayer membrane systems (Naitoh *et al.*, 1991) and in the vesicles of spherical shape (Fendler and Tundo, 1984; Bader *et al.*, 1985). This "organized polymerization" of amphiphilic monomers is a new concept that does not appear in the conventional polymer chemistry literature but is expected to open a new scientific field.

c. Molecular Electronics and LB Films

Studies on the monolayers and LB films of amphiphilic molecules had always been a quiet field of surface science until the 1970s. But since the early 1980s this field has suddenly been in the limelight because of a new concept: "molecular electronics." In the electronic industry, the size of electronic circuits keeps getting smaller and smaller very rapidly: The size of lithography is now measured in submicrometers. If this trend of size-reduction of the LSI is maintained, the size must be in the order of molecules by 2015 to 2020. The concept of molecular electronics appeared from such requirements in the electronic industry, and LB films are expected to be the most possible technique to realize these requirements. Electronic circuits of molecular size of course cannot be made with conventional lithographic techniques and semiconductor wafers. A completely novel technology to assemble the molecular devices as wanted is necessary, and at present the LB method is the only technique available to arrange the molecules.

There are two approaches in the research of the molecular electronics: One is targeted to the molecular circuits and the other is to use the ultra-thin LB films as an electronic device. The former approach must overcome many unbelievably difficult problems: a lot of molecules having the function of unit device must be assembled, such molecules must be assembled on a solid surface in a precise way, and interfaces between the molecular circuits and the keys that computer users touch must be developed. The LB method is good for controlling the arrangement of molecular layers, but quite poor for making the molecular circuits in one LB layer. So this approach is just a dream, not really based on reality. On the other hand, the latter approach is based on the realistic studies. Research has been done to make clear the contribution of an ultra-thin film of molecular size to the electronic devices. As already mentioned previously, the application of the film to an insulator and a resist film for lithography (etc.) has been studied. This approach is to use the LB method as a kind of coating technique of ultra-thin films and thus is the most realistic one for electronic applications.

Refreshing Room!

PHASE SEPARATION AND MASS ACTION MODEL OF MICELLES

The phase-separation model of micelles was used in Sections 3.3.1.*a* and 3.3.2.*a* to obtain the physical meaning of the Krafft point. This model leads necessarily to the interpretation of the Krafft point as a melting temperature. The melting-point model of the Krafft point is very useful for gaining insights as to how to depress the Krafft point (as mentioned in Section 3.3.1.*d*).

The mass action model for micelle formation is the older model (Murray and Hartley, 1935; Moroi, 1992). Micellization is treated in this model as an association of surfactant molecules, like dimer and trimer formation. Accordingly,

$$nm \leftrightarrow M_n \text{ and } K = \frac{[M_n]}{[m]^n}, \tag{3.21}$$

where m and $[m]$ denote the monomer of a surfactant and its concentration, respectively; M_n and $[M_n]$ are the micelle-containing n monomers and its concentration of the agent, respectively; and K is the equilibrium constant. The number of association, n, is considerably large (~50–100), and the concentration of micelle increases steeply with the increasing concentration of monomer. According to this model, the solubility curve is actually BAC rather than BAD in Figure 3.26. This mass action model can also explain the abnormal behavior in the solubility-temperature curve of a surfactant, but it cannot provide any insights for predicting some solution properties of the agent.

We do not have any decisive data to deny either model—readers can use whichever they like. This author, however, prefers the phase separation model because it gives some insights for practical applications of the surfactants.

Chapter 4 | Applications of Surface Active Substances

Surface active substances are compounds that show strong action on surfaces and interfaces and change their properties profoundly. Surfaces and interfaces are, of course, present everywhere in our daily life and in many kinds of industries. So surface active substances, or surfactants, can be used in application fields of every sort. Except for household detergents, however, the amount of surfactant used in each application field is usually small. The surfactant industry is, therefore, obliged to manufacture its products in small amounts and wide variety.

4.1 Interactions of Surfactants and Their Synergistic Effects

Surfactants are very often used in mixed systems to obtain some desired performance of products. In such cases, many kinds of blending effects occur. These blending effects may be classified into three categories:

1. *Complementary effect*:

$$\alpha(a + b + \cdots) = \alpha(a) + \alpha(b) + \cdots$$

or

$$\alpha + \beta + \gamma + \cdots = \alpha(a) + \beta(b) + \gamma(c) + \cdots$$

Here, either the performance of α in a blended product is obtained as a sum of the performance α of several components or several performances are acquired from several compounds, each of which has each performance. In both cases, the blending effect is just the additive effect of each component.

2. *Quantitative synergistic effect*:

$$\alpha(a + b + \cdots) > \alpha(a) + \alpha(b) + \cdots$$

In this effect, a greater performance of α can be obtained than the sum of the performance α of each component by the combination of several components.

3. *Qualitative synergistic effect*:

$$\varphi(a + b + \cdots) \leftarrow \alpha(a) + \beta(b) + \cdots$$

Here, a completely new performance φ appears by blending the several components, each of which has the performance α, β, ...

We tend to expect that interactions occur between each component when the above blending effects (synergistic effects in particular) are obtained. But such an expectation is wrong. The synergistic effect can be obtained, as expected, when some interaction is present between surfactant components, but the opposite proposition is not necessarily true. The synergistic effects on blending some kinds of surfactants will be discussed in this section, taking into consideration the relationship between the synergistic effect and the interaction between the surfactants.

4.1.1 THERMODYNAMIC REPRESENTATION OF SURFACTANT INTERACTIONS

The chemical potential of component A in a solution can be expressed as

$$\mu_A = \mu_A^0 + RT \ln X_A, \tag{4.1}$$

where X_A is the mole fraction and μ_A^0 is the standard chemical potential of component A. R and T are the gas constant and the absolute temperature, respectively. The above representation of the chemical potential is written under the assumption of ideal mixing of component A and the other components, i.e., there is no interaction between them. More strictly speaking, the ideal solution is defined as the athermal mixing that occurs when the interaction energy (W_{AB}) between components A and B can be given as an arithmetic mean of the interactions between the A molecules themselves (W_{AA}) and the B molecules themselves (W_{BB}), i.e., $W_{AB} = (W_{AA} + W_{BB})/2$. Therefore, the deviation (W) from the above arithmetic mean is defined as the extent of interaction energy of real solution of the components A and B:

$$2W = W_{AA} + W_{BB} - 2W_{AB}. \tag{4.2}$$

We can rewrite Eq. (4.1) as Eq. (4.3) when the interaction energy, W, is present between A and B. Thus, we have

$$\mu_A = \mu_A^0 + RT \ln a_A \equiv \mu_A^0 + RT \ln \gamma_A X_A, \tag{4.3}$$

where a_A is the activity (effective concentration) and γ_A is the activity coefficient of the component A. The activity coefficient γ_A is added to Eq. (4.1) because of the presence of the interaction energy W, so that the value of γ_A must be correlated with W. Eq. (4.4) provides this relationship:

$$RT \ln \gamma_A = ZN_A WX_B^2, \tag{4.4}$$

where Z is the coordination number of B molecules surrounding a molecule A and N_A and X_B are the Avogadro's number and the mole fraction of component B, respectively. The solution that satisfies the above relation is called the *regular solution* to result from a relatively small interaction energy W. When the attractive interaction between the A and B molecules is greater than either the A–A or B–B interaction, W must be negative and $\gamma_A < 1$ from Eq. (4.4). The inequality of $\gamma_A < 1$ means that the activity a_A (effective concentration of component A) is less than the real concentration X_A. The situation is opposite, of course, when W is positive and $\gamma_A > 1$.

Figure 4.1 shows the effective concentration a_A plotted against the real com-

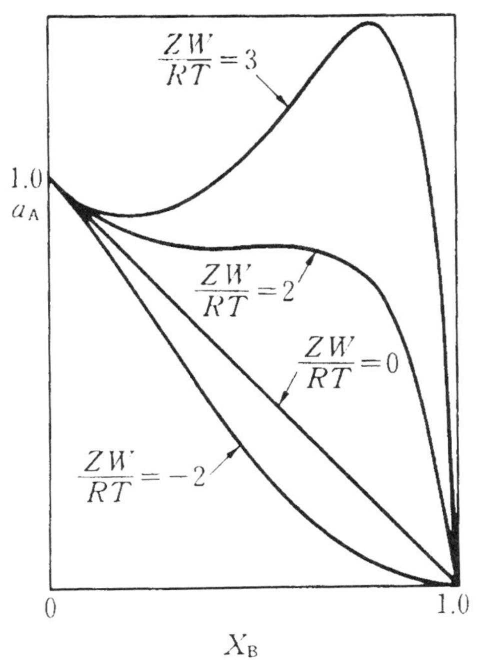

Figure 4.1 Activity coefficient of component A plotted against the mole fraction of component B in various cases of interaction energy. W in this figure corresponds to $N_A W$ in the text.

position of X_B. We can see from the figure that a_A is greater than unity in some concentration range when W is positive and larger than a certain value. It is impossible for the effective concentration (mole fraction) of a_A to be greater than unity; phase separation takes place in such a case. The boundary condition of W for phase separation is $ZN_AW = 2RT$.

Let us consider a special case where the interaction energy W is a very large negative value. The strong attractive interaction between the A and B molecules makes them always form A–B pair molecules in the solution. This A–B pair molecule is usually called an *addition compound* or *complex.* If an addition compound of 1:1 stoichiometry is formed, only the new A–B compound exists at that composition—i.e., no free A or B molecules are present. Consequently, the effective concentration a_A is zero, i.e., $\gamma_A = 0$, and $W = -\infty$ from Eq. (4.4). The infinite interaction energy ($W = -\infty$) is nonsense, of course, since the regular solution theory (Eq. (4.4)) cannot be applied to such a system of strong interaction. But one can understand qualitatively that the W is such a very large negative value in such a case that some addition compound or complex is formed due to the strong interaction between the solution components.

4.1.2 SYNERGISTIC EFFECT IN NONINTERACTING SYSTEMS BETWEEN SURFACTANTS

The stability of surfactant products in the liquid state highly depends on the Krafft point of the agents. Below the Krafft point, surfactant crystals are precipitated out of the liquid state and its surface activity is lost. As a result, the surfactant cannot perform its function below the Krafft point. In addition, the precipitated crystals damage to the appearance of the products. Thus, mixed surfactant systems are frequently used in the formulas of liquid products to depress the Krafft point. Figure 3.29 in Chapter 3 is a typical example (Tsujii *et al.*, 1980): It shows the blending effect of two surfactant components and that the Krafft point of the surfactant mixture is lower than that of either individual surfactant. We must take note that this phenomenon occurs without any interaction between two components of the surfactant.

As already pointed out in Section 3.3.1.*c*, this Krafft point change can be written quantitatively (Eq. (3.16)) as a melting point depression phenomenon. It is important to note that the left-hand side of Eq. (3.16) is represented by only one variable of X_A (not including the activity coefficient γ_A). This means that the Krafft point depression is governed by only the mixing entropy of two surfactant components, and the interaction term between them does not contribute essentially to the Krafft point change. The term of γ_A, of course, appears in the left-hand side of the equation when the interaction W is present, but this term affects only the extent of Krafft point depression and is not essential for the melting (Krafft) point depression phenomenon itself. Eq. (3.16) indicates that the Krafft point is depressed when the mole fraction of surfactant component A is decreased in the micellar phase. This reduction of the mole fraction of surfactant A can be also achieved by solubilization of some organic compounds into micelles; the Krafft point is actually depressed by the addition of fatty alcohols.

In the mixed systems of ionic surfactants, the Krafft point can be decreased also in terms of the mixing (entropy gain) of counterions. To show this, we modify the equation of Krafft point depression (Eq. (3.16)) as

$$\nu_{A^+} \ln X_{A^+} + \nu_{A^-} \ln X_{A^-} = -\frac{\Delta H_A^0}{R}\left(\frac{1}{T} - \frac{1}{T^0}\right), \tag{4.5}$$

where ν_{A+} and ν_{A-} are, respectively, the number of positive (counter-) and negative (surface active) ions in one surfactant molecule. The first and second terms in the left-hand side of Eq. (4.5) express the contribution of the mixing of the counterions and the surface active ions, respectively, to the Krafft point depression. Figure 4.2 shows the plots of $\ln X_{\mathrm{Na}}$ or $\ln X_{\mathrm{Ca}}$ as a function of $1/T$ in the binary mixture of sodium and calcium dodecyloxyethylene sulfate (Tsujii *et al.*, 1980, and see Section 3.3.1.*c*). Good straight lines were obtained, and the slopes of these plots gave us reasonable enthalpy values of fusion for the above two surfactants.

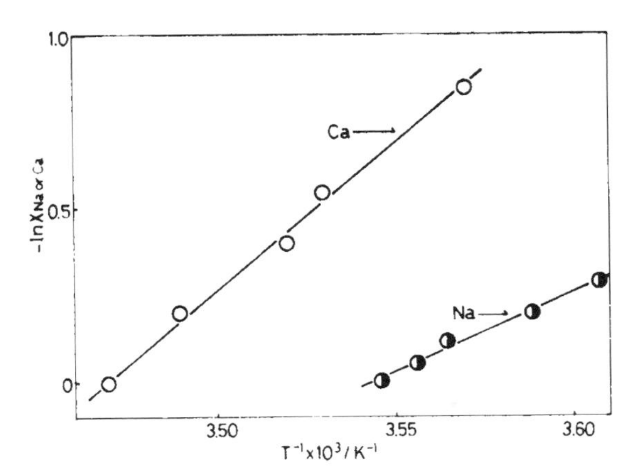

Figure 4.2 The plots of $\ln X_{\mathrm{Na}}$ or $\ln X_{\mathrm{Ca}}$ as a function of $1/T$ in the binary mixture of sodium and calcium dodecyloxyethylene sulfate using the data shown in Figure 3.29 (Tsujii *et al.*, 1980). Reprinted with permission from *J. Phys. Chem.*, **84,** 2287 (1980), American Chemical Society.

We may be able to take the HLB adjustment in nonionic surfactant mixtures to obtain the stable emulsions as one more example of the synergistic effect with no interaction. A remarkable synergistic effect is obtained from the viewpoint of emulsion stability, but the interaction energy between the nonionic surfactant molecules of similar chemical structure must be small enough.

4.1.3 SYNERGISTIC EFFECT CORRELATED WITH THE EXTENT OF INTERACTION BETWEEN SURFACTANTS

The critical micelle concentration (CMC) is the only solution property of surfactants whose synergistic effect is quantitatively correlated with the extent of interactions between components. The CMC values in binary surfactant mixtures with no interaction between them are shown first in Figure 4.3 as a reference

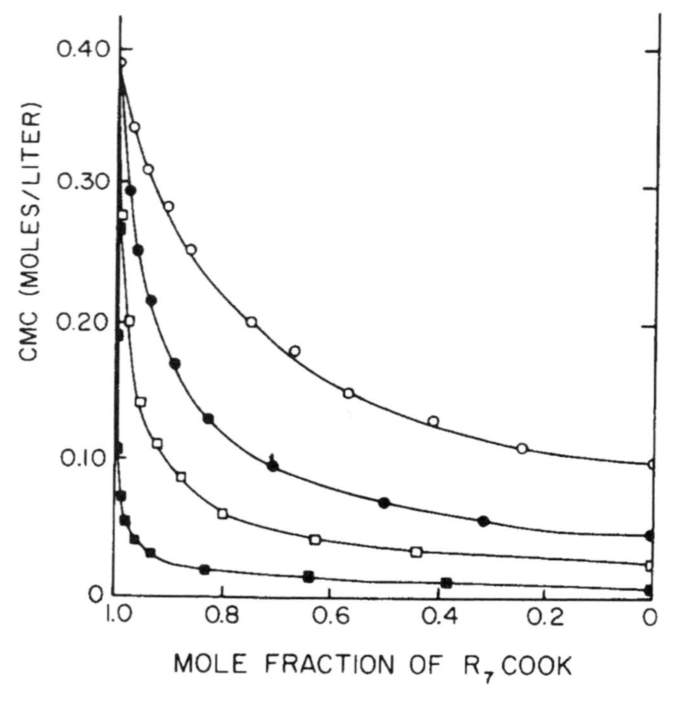

Figure 4.3 CMC values of binary surfactant systems of ideal mixing. Data curves are the mixtures of potassium octanoate + (from top to bottom) potassium decanoate, undecanoate, dodecanoate, and tetradecanoate (Shinoda, 1963).

(Shinoda, 1963). The CMCs of mixed systems are in between the CMC value of each surfactant component and change monotonously with varying composition. This CMC-composition curve is similar to the vapor-phase curve in the phase diagram of liquid/vapor equilibrium of ideal solutions. The curve deviates from the ideal when two components interact with each other. If the interaction is strong enough, the CMC values of the mixed surfactant deviate much and become even lower or higher than either value of each component; Figure 4.4 shows an example of a mixed system that exhibits the lower CMC values (Lange and Beck, 1973). This phenomenon occurs when two components interact attractively with each other ($W < 0$).

The interaction parameters, $\beta(=ZN_A W/RT)$, for various combinations of surfactant are listed in Table 4.1. All values of the parameter listed in the table are negative, and these mixed systems show lower CMC values than those of ideal ones. The second and third systems from the bottom are anionic/cationic and anionic/amphoteric ones, respectively, and give particularly large negative

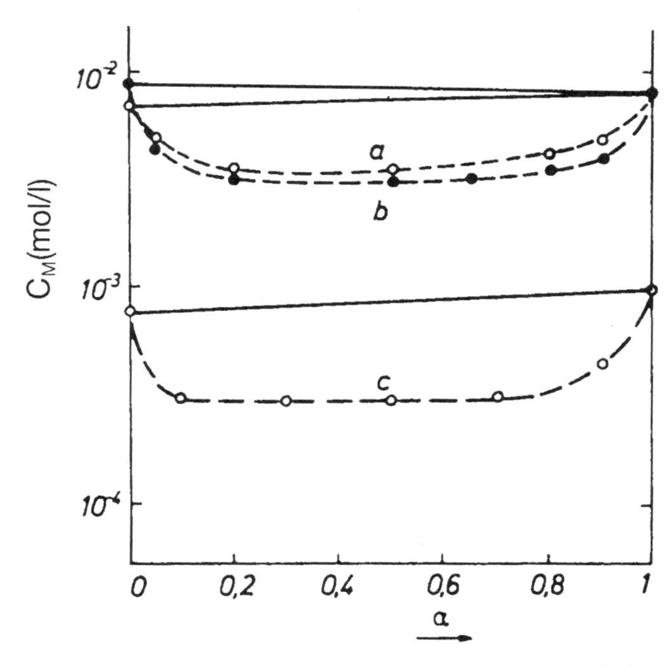

Figure 4.4 The CMC of mixed surfactant systems of alkylpolyoxyethylene ether and alkyl sulfate (Lange and Beck, 1973). Shown here are the $C_8(EO)_4$-C_{12}-sulfate (a), $C_8(EO)_{12}$-C_{12}-sulfate (b), and $C_{10}(EO)_6$-C_{15}-sulfate (c) systems.

Table 4.1 Interaction Parameters (β) of Various Surfactant Mixtures

Mixed System	*β*
$C_{12}OSO_3Na/C_8(EO)_6$	–4.1
$C_{12}OSO_3Na/C_{12}(EO)_8$	–3.9
$C_{15}OSO_3Na/C_{10}(EO)_6$	–4.3
$C_{12}\phi SO_3Na^*/C_{12}NO(CH_3)_2$	–3.5
$C_{16}N(CH_3)_3Cl/C_{12}(EO)_5$	–2.4
$C_{14}N(CH_3)_3Cl/C_{10}(EO)_5$	–1.5
$C_{14}N(CH_3)_2\text{-}CH_2\phi Cl^*/C_{10}(EO)_5$	–1.5
$C_{20}N(CH_3)_3Cl/C_{12}(EO)_8$	–4.6
$C_{10}P(CH_3)_2O/C_{10}S(CH_3)O$	–0.84
$C_{12}N(CH_3)_3Cl/C_{12}N(CH_3)_2(CH_2)_3SO_3$	–1.0
$C_{12}OSO_3Na/C_{12}N(CH_3)_2(CH_2)_3SO_3$	–7.8
$C_{10}OSO_3Na/C_{10}N(CH_3)_3Br$	–18.5
$C_{12}P(CH_3)_2O/C_{12}N(CH_3)_2(CH_2)_5COO$	–1.0

*ϕ: Phenyl group

values. They form the addition compound discussed in the next section, and their interaction may be too strong for the regular solution theory to be applied.

The opposite cases, i.e., mixed systems with higher CMC values than either component are demonstrated in Figure 4.5 (Mukerjee and Yang, 1976). These are mixtures of surfactants having a fluorocarbon and a hydrocarbon chain as a hydrophobic group. The interaction energy between the fluorocarbon and the hydrocarbon chains is remarkably small (W is a large positive value) in these systems. In a mixed system where the interaction energy (W) is larger still, these two kinds of surfactant—fluorocarbon and the hydrocarbon chain—are no longer miscible with each other. In such cases the micelles are phase-separated, and two kinds of micelles—the fluorocarbon surfactant- and hydrocarbon surfactant-rich—are present at the same time in a solution (Shinoda and Nomura, 1980; Funasaki and Hada, 1980; Funasaki, 1993).

4.1.4 ADDITION COMPOUND (COMPLEX) FORMATION AND SYNERGISTIC EFFECTS

In this section, we will discuss mixed surfactant systems in which two components interact most strongly to form an addition compound or complex. Very strong synergistic effects can be expected in these systems since a new com-

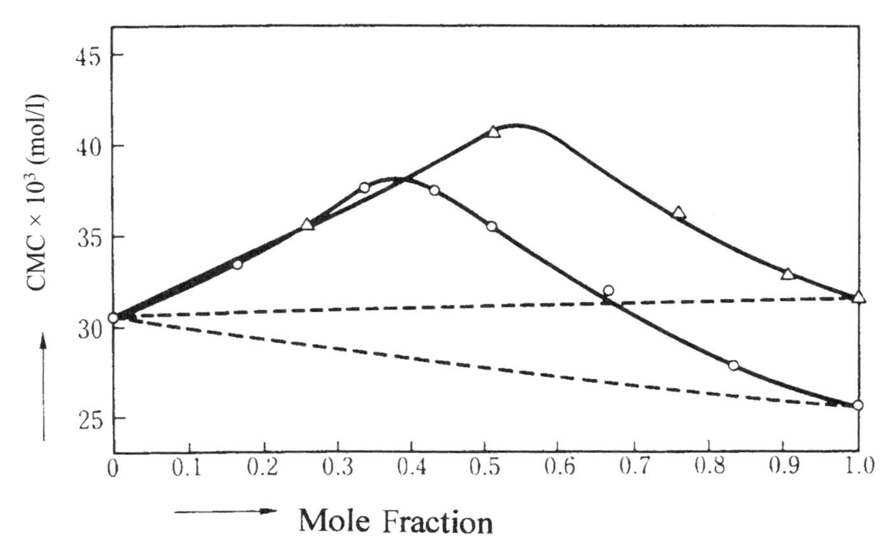

Figure 4.5 CMC values of mixed systems of fluorocarbon- and hydrocarbon-type surfactant. The CMC curves are for the sodium perfluorooctanoate-sodium dodecanoate (○) and sodium perfluorooctanoate-sodium decyl sulfate (△) systems (modified from Mukerjee and Yang, 1976). Reprinted with permission from *J. Phys. Chem.*, **80,** 1388 (1976), American Chemical Society.

pound is formed from the two kinds of component. Several typical synergistic systems are indeed classified into this mixed system.

Amphoteric surfactants are used in most cases as a cosurfactant for anionic agents in shampoos and/or liquid detergents because of their boosting effect in the foaming property and the detergency of anionic surfactants. This boosting effect is the most typical example of a synergistic effect originating from the addition compound formation. The Krafft point maxima caused by the addition compound formation were shown in the binary mixtures of sodium dodecyl sulfate and amphoteric alkylsulfobetaines (Figure 3.32). Figure 4.6 exhibits another example of addition compound formation between anionic sodium alkyl sulfates and amphoteric N-dodecyl-β-alanine (Tajima *et al.*, 1979). The surface tension of the mixed system decreases to much lower concentrations than that of each individual component, which indicates that the complex is much more surface active. An addition compound formed between the two surfactants in Figure 4.6 can be regarded as a kind of double-chain surfactant and may show higher surface activity.

Some other examples of addition compound formation have been known in

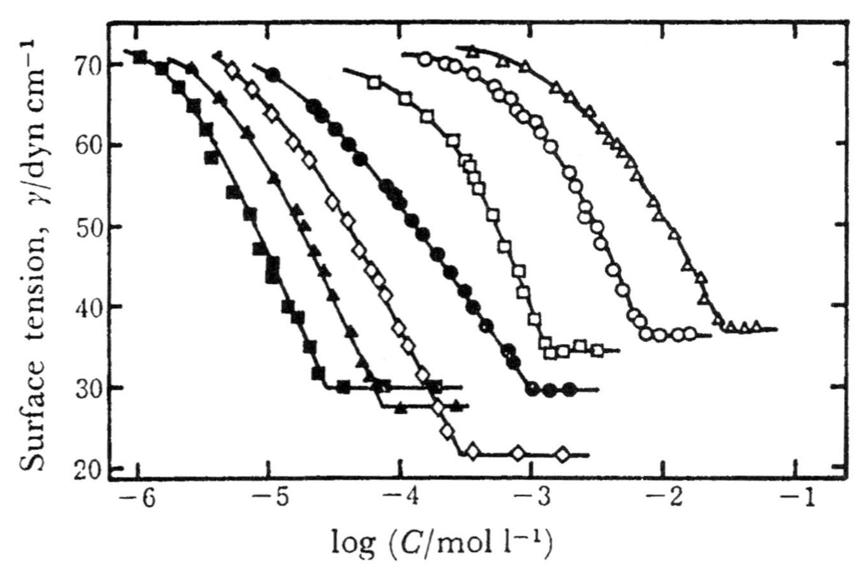

Figure 4.6 Surface tension vs. concentration curves of the binary mixtures of sodium alkyl sulfates and amphoteric N-dodecyl-β-alanine as well as those of their single solutions (Tajima *et al.*, 1979). Single solutions: sodium decyl sulfate (△), sodium dodecyl sulfate (○), sodium tetradecyl sulfate (□), and N-dodecyl-β-alanine (•). Equimolar mixtures: N-dodecyl-β-alanine with sodium decyl sulfate (◇), with dodecyl sulfate (▲), and with tetradecyl sulfate (■). Reprinted with permission of the Chemical Society of Japan.

the binary mixtures of soap/fatty alcohol (Shah, 1970), alkyl sulfate/fatty alcohol (Kung and Goddard, 1963), and soap/fatty acid (Lucassen, 1966; Feinstein and Rosano, 1969; Mino *et al.*, 1976). The mixed systems of soap/fatty alcohol and soap/fatty acid are often used practically in toiletry soaps with creamy foams. The binding of surfactants onto some water-soluble polymer chains in aqueous solution is also a kind of complex formation (Goddard, 1990; Goddard and Ananthapadmanabhan, 1993). This polymer-surfactant complex is sometimes applied to shampoos, liquid detergents, and so on.

4.2 Detergency and Surface Activity

The fundamental properties and functions of surfactants can be applied practically to various fields. Adsorption and aggregation were discussed in Section

3.1, and applications of these functions will be mentioned in detail later. Here, we discuss the detergency function separately from the other properties because detergency is a very complex action that cannot be understood by only a single action of adsorption or aggregation of the surfactant. In addition, the detergency is the most widespread and the chief application of surfactants and is worth taking up as a single section.

4.2.1 WASHING PROCESS AND SURFACE ACTIVITY

A surfactant is often called *detergent,* which implies that the most typical and main application of surfactants is the detergent. Applicability of some typical surfactants to various kinds of detergent is shown in Table 4.2. The surfactant, of course, possesses some but not all functions required for detergency. Thus, a detergent is formulated with many kinds of components called *builders* in addition to the surfactant to obtain enough detergency. First we will make clear the actions necessary for detergency; then we will discuss them in relation with the functions and properties of the detergent components.

a. Wetting Process

The detergent process starts with wetting since washing is usually done in water and the soiled material must be in contact with a detergent solution. For example, a cloth soiled with oily dirt is hydrophobic; it floats on the water. It thus is not in contact with the water and so cannot be washed until it is wetted. The surfactant components in a detergent accelerate the wetting of the cloth with water and start up the washing process. As mentioned in Section 3.2.1.*c*, the surfactant adsorption enhances all parts of the wetting phenomena: adhesion, spreading, and immersion. The adhesion and the spreading wetting are helpful for washing of flat solid surfaces like dishes, metals, etc., and the immersion wetting is needed for porous materials like clothes.

b. Removing Process of Soils

Removing soils from a substrate is the central process of cleansing. This process is the detergent process itself in a narrow sense. The removing of soils is the reverse process of coagulation or adhesion, and it requires energy. The soils must be transferred from the attached state (attached to a substrate) of a lower energy level to the separated state in an aqueous medium of a higher energy level. This energy is supplied by an external source, such as a washing machine, a ultra-son-

Table 4.2 Applicability of Some Typical Surfactants to Detergents

	Surfactant	Natural or Petroleum	Heavy-duty Powder	Heavy-duty Liquid	Dishwashing	Shampoo	Body-Cleanser (for skin)	House Room	Fabric Softener/ Hair Conditioner
ANIONICS	Alkylbenzene sulfonate (ABS, LAS)	P	◎	◎	◎			○	
	α-olefin sulfonate (AOS)	P	◎	○	◎	△		△	
	Alkane sulfonate (SAS)	P	△		△			△	
	Alkyl sulfate (AS)	N, P	○	○	○	◎	△	○	
	Alkylpolyoxyethylene sulfate (AES)	N, P	◎	◎	◎	◎	△	○	
	Alkyl phosphate (MAP)	N				△	◎		
	Soap (Fatty acid salt)	N	○	△	△		◎	△	
	α-sulfo-fatty acid ester	N	△						
CATIONICS	Alkyltrimethyl ammonium salt	N		△					◎
	Dialkyldimethyl ammonium salt	N		△					◎
	Alkyldimethylbenzyl ammonium salt	N							
NONIONICS	Alkylpolyoxythylene ether	N, P	◎	○	△			○	△
	Fatty acid diethanolamide	N			○	○	△	△	
	Alkyl polyglucoside	N			◎	○	△	△	
AMPHOTERICS	Alkyldimethylamine oxide	N			○	△		△	
	N-alkyl carboxybetaine	N			○	○			
	Alkyl imidazoline derivatives	N				◎	△		△

◎:frequently used as a main component, ○: sometimes used as a main component or frequently used as a cosurfactant or booster, △: rarely used, No mark: not used.

icator, or a human hand. In this energy-required process of cleansing, the detergent serves to reduce the amount of energy required and to remove the soils easily.

The soil-removing process for solid soil particles was formulated first by Lange as a reverse process of coagulation (Lange, 1967). The DLVO theory discussed in Section 3.2.4.*b* is a main tool for understanding this reverse process. Figure 4.7 illustrates the mechanism for removing solid particles from a substrate that Lange proposed. A soil particle P attached to a substrate S is spontaneously separated at a tiny distance δ in detergent solutions. This step 1 proceeds by the adsorption of surfactant or water molecules onto the soil and/or the substrate surface, and the separated distance δ is in a molecular order. The energy required in step 1 ($A_1 = V_{II} - V_I$) is given spontaneously by the free energy gain of the adsorption process; it need not be supplied externally. Step 2 is to carry the soil particle away from the distance δ to an infinite one. The energy A_2 ($= V_{III} - V_{II}$) is necessary in Step 2 as a sum, but a higher energy barrier E becomes present in the process of this step and must be overcome. Figure 4.8 shows the potential energy vs. distance curve for these steps. One can understand from the figure that the total energy $A_2 + E$ should be supplied externally for Step 2 and is the criterion that determines the ease of cleansing. In Section 3.2.4.*b*, the process from the right to the left side in Figure 4.8 (approaching process of two particles) was taken into consideration to discuss the colloid stability in relation to the potential barrier E. In the cleansing process, on the other hand, the process left to right is important.

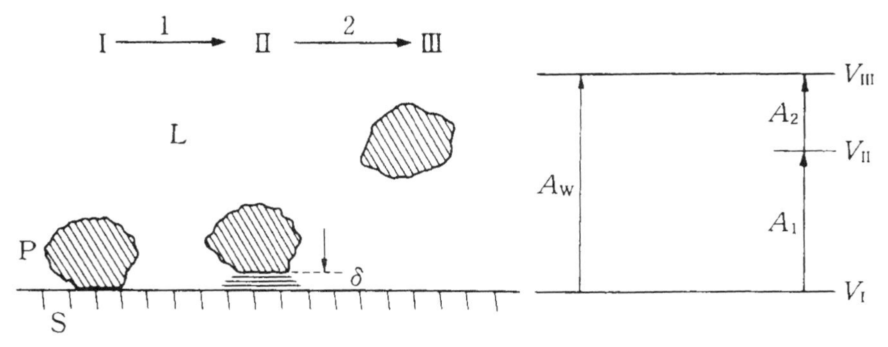

Figure 4.7 Schematic representation of the removing process of a solid (soil) particle (P) from a substrate (S). The energy changes in this process are given in the right-hand figure (modified from Lange, 1967). Reprinted from "Solvent Properties of Surfactant Solutions," p. 126 by courtesy of Marcel Dekker, Inc.

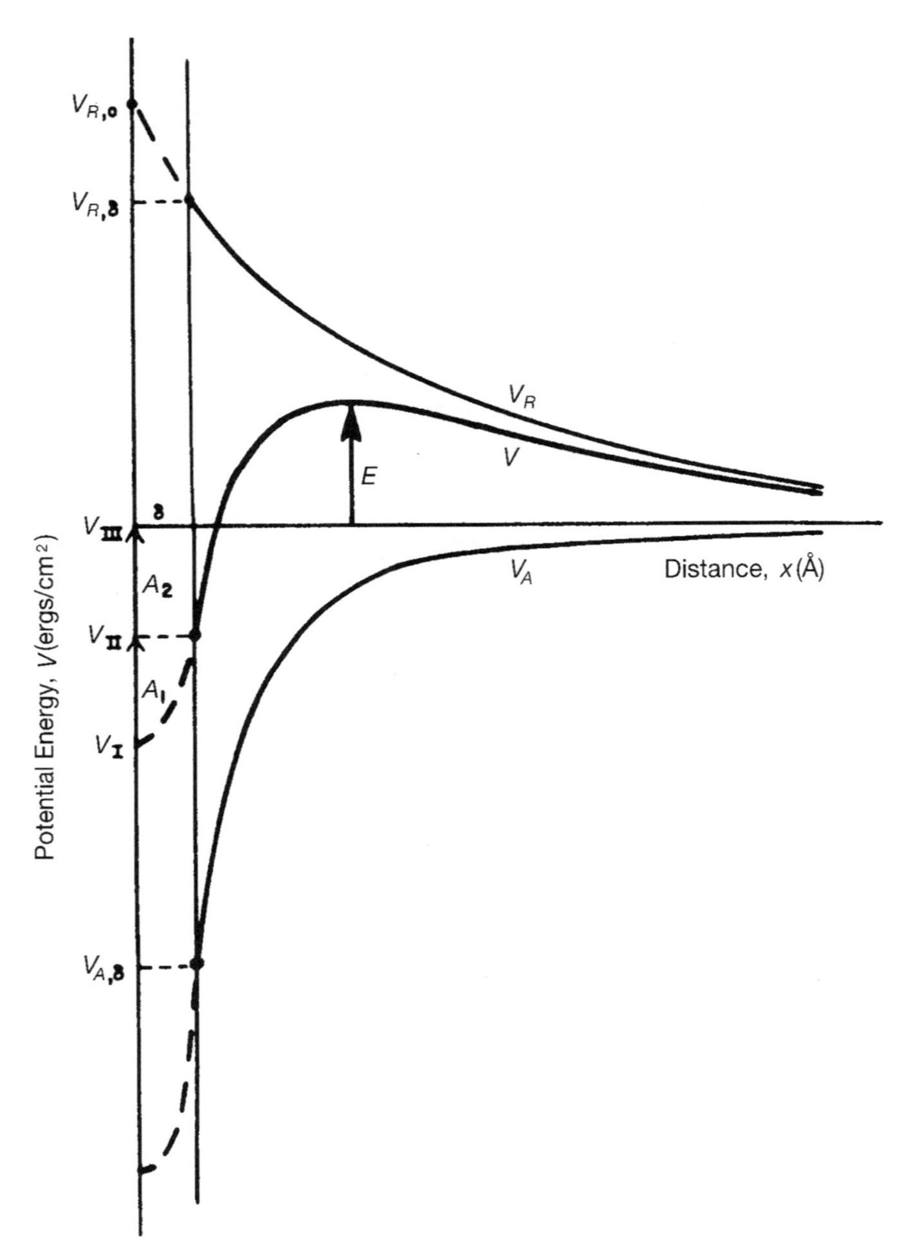

Figure 4.8 Potential energy as a function of the distance between a soil particle and a substrate (Lange, 1967). The soil particle must go over the potential barrier (A_2 + $\boldsymbol{E}$) in the removing process of the particle. Reprinted from "Solvent Properties of Surfactant Solutions," p. 128 by courtesy of Marcel Dekker, Inc.

A detergent component, a surfactant in particular, changes the potential curve in Figure 4.8 and affects the detergency. Anionic surfactants (most commonly used in detergent formulations) adsorb on the soil particles and the substrate, and then give them negative charges in aqueous detergent solutions. Figure 4.9 demonstrates the effect of surface potential (i.e., surface charge) on the potential energy curve of soil release. The value of Z in the figure is proportional to the surface potential ψ_0 and is equal to $ve\psi_0/kT$, where v and e are the surface charge density and the elementary charge, respectively. The figure shows that the potential energy barrier from left to right becomes smaller and from right to left becomes higher with increasing ψ_0. This change in the potential energy curve means that the soils are easily removed from and not easily redeposited onto the substrate; this of course is favorable in the cleansing process. One can easily imagine intuitively that the accumulated negative charges on both particles and substrates in terms of the adsorption of surfactant and/or anionic polymers enhance the release of soils and protect from redeposition owing to their electrostatic repulsion. Since the solid soil particles and the substrates generally bear the negative charges in an aqueous solution, it is advantageous to use anionic surfactants in detergent systems to put more negative charges to them. This is why anionic surfactants are used most often in detergent formulations and cationic ones are not.

Let us move on to the removal of oily (liquid) dirt. Figure 4.10 illustrates an attached oil droplet to a substrate surface. As mentioned in Section 3.2.1.*c*, three interfacial tensions are balanced at one point at which three phases of oil, water, and substrate are in contact with each other (Young's equation):

$$\gamma_{WS} = \gamma_{OS} + \gamma_{WO} \cos \theta. \tag{4.6}$$

Here γ_{WS}, γ_{OS}, and γ_{WO} are the interfacial tensions of water/substrate, oil/substrate, and water/oil interfaces, respectively, and θ is the contact angle between the oil and substrate. What happens on this oil droplet when surfactant is added to the water phase? The surfactant molecules adsorb on the interfaces of water/oil and water/substrate, making their interfacial tensions lower. The interfacial tension of oil/substrate, however, does not change since no surfactant molecules can adsorb on it. (Usually, anionic surfactants are not oil-soluble.) Thus, the contact point of the oil/water/substrate phases is pulled by the relatively large interfacial tension of γ_{OS}, and the oil droplet becomes more round. This deformation in the shape of the oil droplet is called *rolling-up*.

Figure 4.11 shows schematically the rolling-up process. If this phenomenon takes place in the extreme, the interfacial tensions of γ_{WS} and γ_{WO} becomes almost zero and the oil becomes spherical. In such a case the contact angle is al-

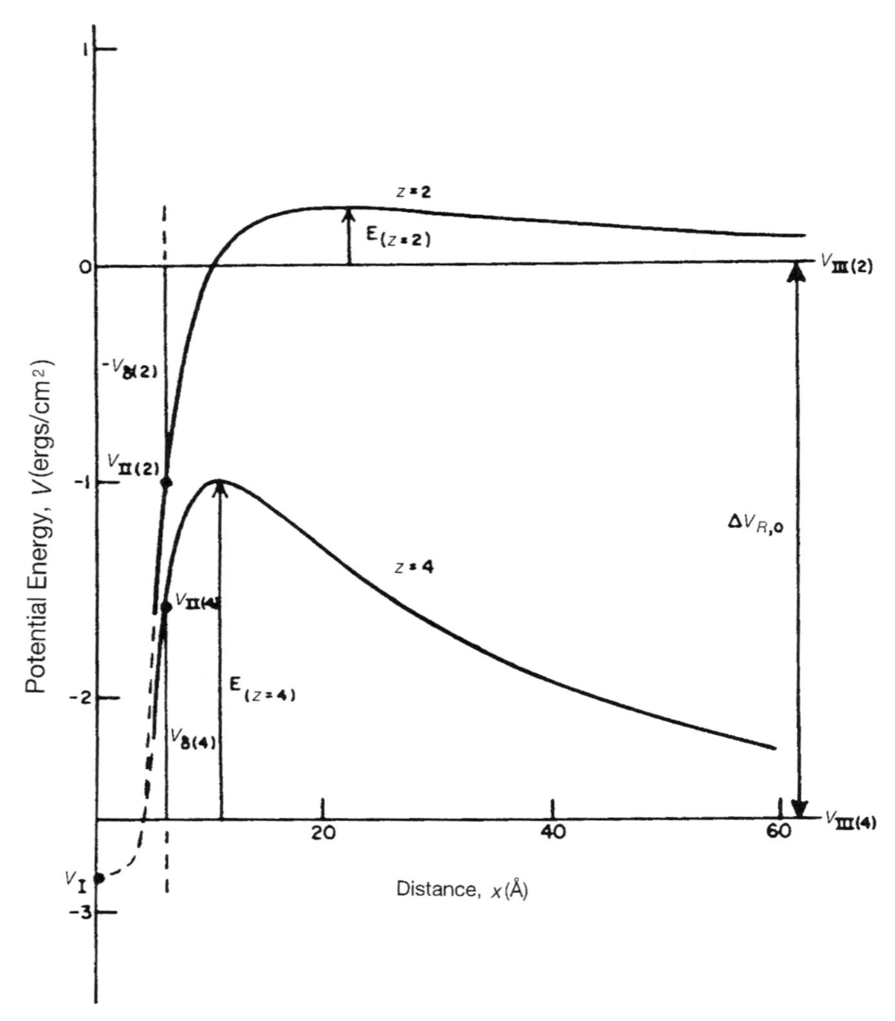

Figure 4.9 Effect of surface charge ($Z = ve\psi_0/kT$) on the potential energy between a soil particle and a substrate. The soil is easily removed and not easily redeposited on the substrate with increasing surface charge (Lange, 1967). Reprinted from "Solvent Properties of Surfactant Solutions," p. 145 by courtesy of Marcel Dekker, Inc.

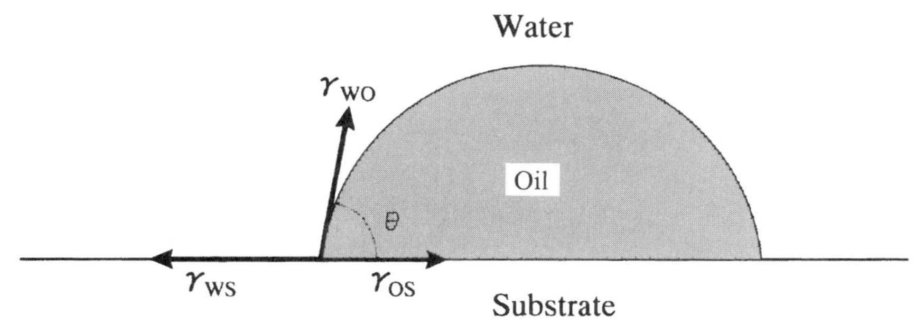

Figure 4.10 An oily dirt attached to a substrate placed in water. The contact angle is determined by the balance of horizontal components of three interfacial tensions.

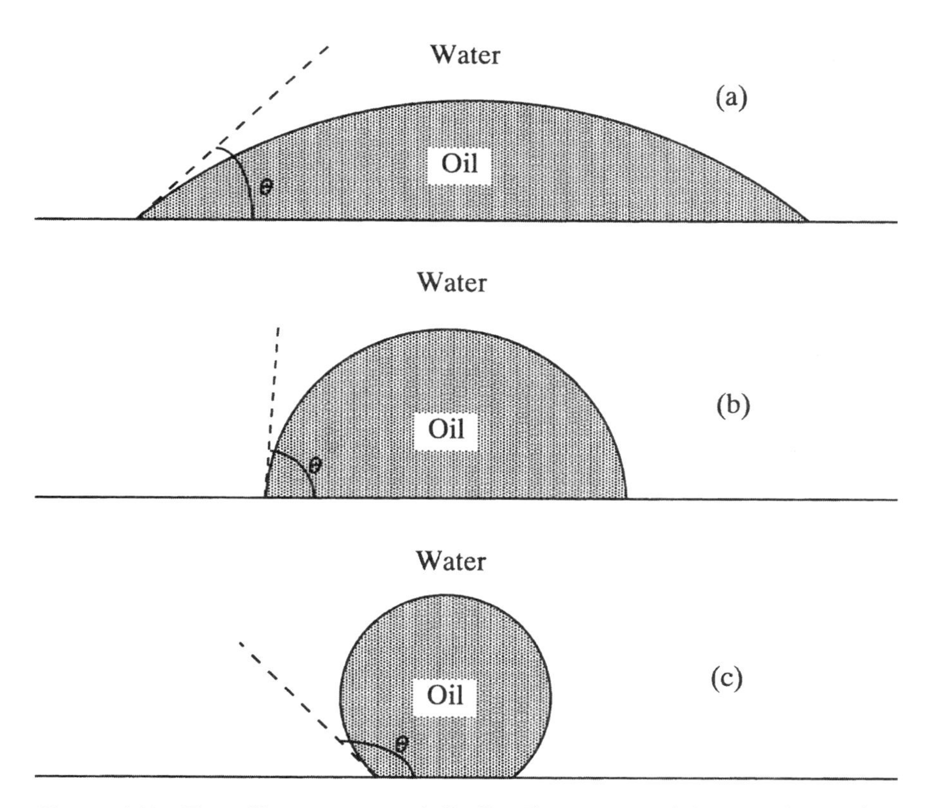

Figure 4.11 The rolling-up process of oily dirt. The contact angle becomes larger when the surfactant is added to the water phase, and the oily dirt can be easily removed. (See text.)

most 180°, and the oily dirt is detached spontaneously from the substrate surface. Even in a case not so extreme, the oily dirt can be easily removed when the contact angle is greater than 90°.

In addition to the rolling-up mechanism, emulsification of oils by a surfactant is also useful in the cleansing process. When oily dirt contains fatty acids—such as soils from the human body, which actually contain fatty acids up to 30%—a special cleansing mechanism works spontaneously. The fatty acids in the soils are neutralized with an alkaline solution of detergent to form soaps. Some soaps are water-soluble and emulsify the oily dirt itself into a bulk aqueous phase. Such a cleansing mechanism also may work in the scouring process of raw wool, cotton, etc.

c. Protection from Redeposition of Soils

Because soils once removed from a substrate can be deposited again onto the substrate, a method is needed to prevent such redeposition. The redeposition of soils is the coagulation process, and the method or technology to prevent it is the same as that to stabilize the colloidal dispersions (see Section 3.2.4).

As understood from the above discussions, detergency is determined by a counterbalance between the removing and the redeposition of soils. The competition of the opposing processes in the potential energy curve of Figure 4.8 governs the detergency. If an equilibrium is attained in this competition, more soil particles will be in the state of the lower energy level, i.e., attached to the substrate. Thus, good detergency is not obtained in equilibrium. This problem may be overcome by taking the space of each energy level into account. The space belonging to the lower energy level (the attached state) is just the vicinity of the substrate surface, i.e., approximately up to the distance of Debye length (~1–100 nm) from the surface. The space of the higher energy level (the removed state) must be almost all cleansing solution. The volume ratio of these spaces is very large, and most of the soil particles could be in the removed state. The number of statistical states in the removed one is very large; thus the washing process may go smoothly even if the Boltzmann distribution is established for the soil particles.

d. Detergent Compositions and Their Roles in Cleansing

So far, we have described the functions and actions required for detergency. Here we will mention the detergent components responsible for these functions.

The surfactant is, of course, the main component and works in several cleansing stages. First, it enhances the wetting of substrates with a cleansing solution. Second, it adsorbs on the soil and substrate surfaces and gives them negative charges, which elevate the surface potential and make the soils release easily from the substrate. Third, it rolls up oily dirt. Fourth, it prevents the soils from redeposition on the substrate in terms of its emulsifying and dispersing functions. Thus, a surfactant is working in almost all stages of the detergent process. Soap is sometimes blended for a special performance. Although soap works as a surfactant in an alkaline solution of the washing step, it works as a defoaming agent in the rinsing step.

Zeolite is a sequestering agent for multivalent metal ions. (Polyphosphates such as sodium tripolyphosphate were used for this purpose until their contribution to the eutrophication problem of lakes and swamps was discovered; now zeolite is used.) Calcium and/or magnesium ions in hard (tap) water form a salt with an anionic surfactant whose Krafft point is much higher than ambient temperature. The surfactant salt of high Krafft point precipitates out of the cleansing solutions, thus losing its surface active properties. The soap scum frequently seen in bathrooms is such a salt. Zeolite sequesters the multivalent ions and prevents the anionic surfactants from precipitating out of the solutions.

Enzymes are now very popularly blended into detergent formulations. Protease and lipase are used to digest the soils of proteins and lipids, respectively. Cellulase has been recently developed as an enzyme component of a detergent that attacks a substrate (cotton) to remove the soils. This cellulase technology, a breakthrough in the detergent industry that makes the detergent volume compact, will be described in detail in the next section.

Water-soluble polymers are used as an agent to protect against redeposition of soils. Carboxymethyl cellulose (cmc) is the most popular antiredeposition agent; sodium polyacrylate is also sometimes used. These anionic polymers adsorb on the soils and substrates and give them negative charges. The steric effect of nonionic polymers on the stability of dispersion (Section 3.2.4.*c*) is also utilized for antiredeposition. Polyethylene glycol is a good dispersant for this purpose.

An alkaline condition in a cleansing solution is useful to give negative charges to the soils and the substrates. Sodium carbonate and sodium silicate are typical examples of such alkaline agents. Oily soils containing fatty acids can be spontaneously removed in alkaline solutions owing to the formation of soaps in the dirt.

Bleaching agents and activators of them are, of course, used to make white clothes clean. These agents provide another key technology for compacting detergent volume and will be discussed in the next section.

4.2.2 HEAVY-DUTY DETERGENTS

Heavy-duty detergent for home laundry use is doubtless the leader among detergents. The basic theory for detergency in the home laundry of clothes is, of course, the same as we have been discussing. But a great innovation has been recently made in this field to compact the detergent volume up to 1/4–1.5. The key technologies that make this possible are those of enzymes and bleaching agents. Let us discuss first the pioneering work of this innovation: the alkaline cellulase technology (Murata *et al.*, 1991; 1993; Hoshino and Ito, 1997).

The most innovative and interesting point of the alkaline cellulase technology is that it changes the concept of the detergent mechanism. The conventional idea for detergency is that the soil itself must be attacked for it to be removed from a substrate. Thus, protease and lipase are formulated to attack the soils of proteins and fats even when enzymes are utilized. In the cellulase technology, on the other hand, the substrate (cotton) is attacked to remove the soils from it. This new idea comes from the precise analysis of attached state of dirt, which revealed that dirt or soils are attached not only on the surface of the cotton fibers but also inside individual fibers (see Figure 4.12). It has been further made clear that the soils are present particularly in the amorphous part of the cellulose fiber. The soils present inside a fiber cannot be removed by conventional detergent components (such as surfactants and builders). To remove the soils inside the fiber, the polymer networks of cellulose molecules in the amorphous part must be partially cleaved to open a space for the detergent components to enter and attack the soils.

A number of cellulases have been tested for detergent applications; Figure 4.13 shows the enzymatic properties of the best one, produced by an alkalophilic *Bacillus* bacterium. The optimum pH of this enzyme is about 10, which is quite suitable for detergent use. The research most intensively done during the above development was to improve the enzyme activity, since the enzyme cost must be low enough to be formulated in detergent. One can understand the excellent detergent effect of this enzyme from Figure 4.12. Other enzymes such as protease and lipase, which are designed to work at high pH ($\approx$10) and in surfactant solutions, also contribute to the compactness of detergents.

One more key technology in detergent compacting is that of the bleaching agent and its activator. Since white clothes such as shirts and underwear are abundant laundry items in the home, the bleaching technology is particularly important for home-use detergents. Perborates and percarbonates have conventionally been used as bleaching agents in detergent, but an activator for them has only recently been developed. Sodium nonanoyloxybenzene sulfonate (NOBS), the activator for sodium perborate, accelerates the oxidation effect.

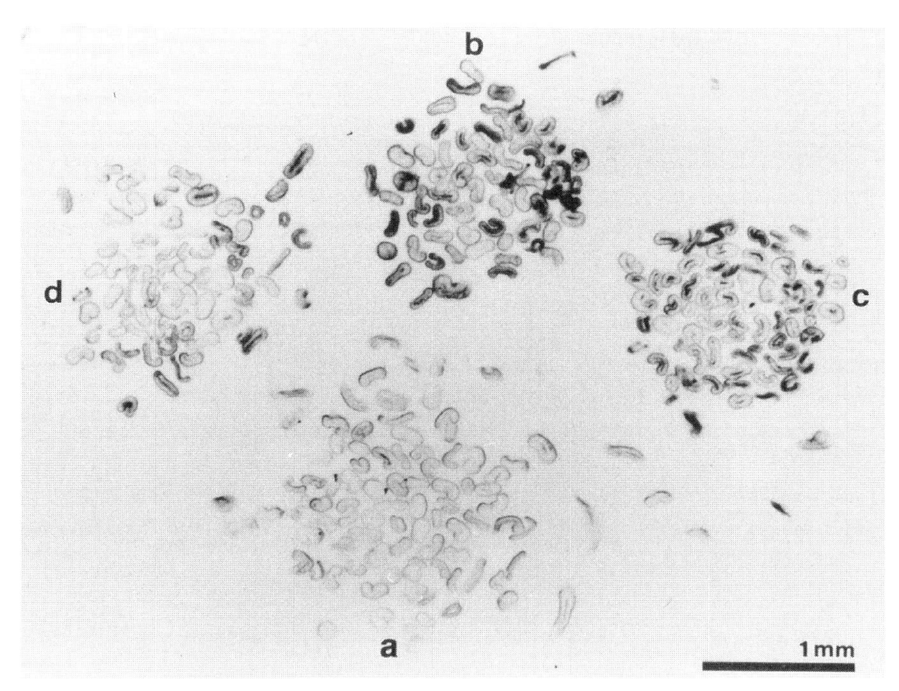

Figure 4.12 Optical microscope photographs of the cross sections of a single cotton fiber stained with osmium tetroxide. Shown here are a yarn of a new cotton undershirt (a), a naturally soiled undershirt (b), an undershirt washed by a detergent containing a protease (c), and an undershirt washed by a detergent containing an alkaline cellulase (Murata *et al.*, 1991).

The surfactants used in heavy-duty detergents are mainly anionic ones such as linear alkylbenzene sulfonates, alkyl sulfates, alkylpolyoxyethylene sulfates, α-olefin sulfonates, and so on. Soaps are sometimes used as a foam-controlling agent. Polyoxyethylene alkyl ether, which is a nonionic surfactant, shows strong detergency for oily dirt. However, this surfactant is a liquid material, which is hard to formulate in powder detergents. A new technology to powder the liquid nonionic surfactant has recently been developed: A silicate powder absorbs a large amount of the nonionic surfactant in the interparticle spaces, making it a dry powder that can be applied to home-use detergents.

Heavy-duty detergents contain a number of components called *builders.* They are zeolite, sodium silicate or carbonate, water-soluble polymers, etc., and their functions were mentioned previously.

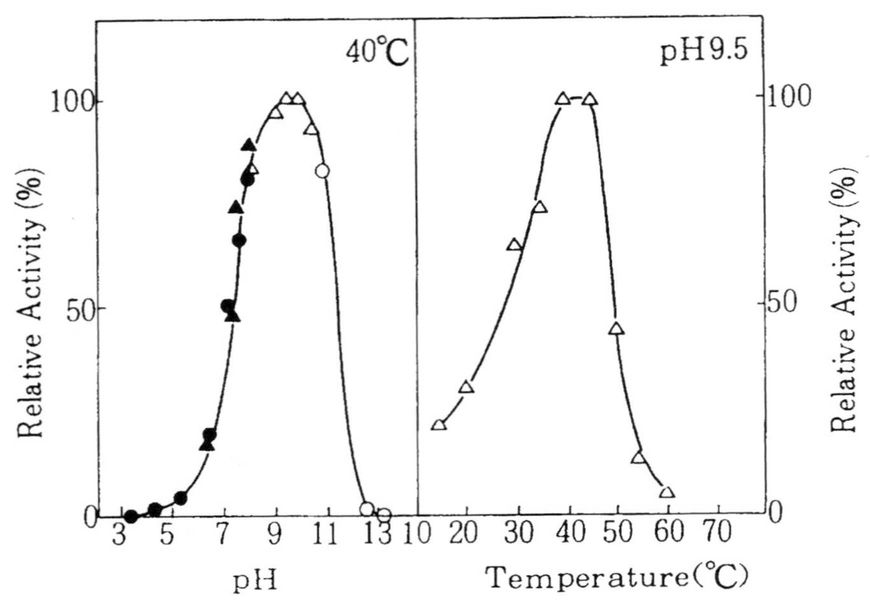

Figure 4.13 Enzymatic properties of the alkaline cellulase used in the compact detergent, optimum pH and temperature (Hoshino, 1991). The different symbols indicate the different buffers used.

4.2.3 CLEANSERS FOR THE HUMAN BODY

Special care must be taken when developing cleansers for the human body. The skin performs the important physiological functions of sensing, secretion, respiration, etc., and cleansers must be harmless for these functions. In fact, cleansing is done to keep these physiological functions normal. Thus, the effects of cleansers and their components upon the human body must always be considered carefully.

a. Shampoos

Shampoo today is almost more of a cosmetic than a cleanser. Because most people shampoo their hair almost every day, the soils or dirt to be washed are quite light. Shampoo is used mainly to wash the hair, but also to cleanse the scalp. Therefore the shampoo components, surfactants in particular, must be mild to the skin. In addition, the surfactants used in shampoos should not give a stiff feeling to the washed hair fibers. The chief surfactants that satisfy these requirements are alkylpoly oxyethylene sulfates, alkyl sulfates, and N-acyl taurate as anionic surfactants; alkyl polyglucoside as nonionics; and imidazoline derivatives as amphoteric ones. Alkylcarboxy betaines and alkanolamide of fatty acids are often used as a booster for the foaming property of anionic surfactants.

Two other important components of shampoos besides surfactants are antidandruff agents and cationic polymers that give a conditioning effect to hair. Zinc Pyrithione (zinc bis-(2-pyridylthio-1-oxide)) and Octopirox (ethanol-ammonium 1,hydroxy-4-methyl-6-(2,4,4 trimethyl pentyl)-2-(1H)-pyridinone) are the most popularly used antidandruff agents, and their chemical structures are shown in Figure 4.14. The conditioning effect of shampoos reduces the squeaky and frictional feeling and gives a smooth washing process. Cationic cellulose is most frequently blended for this purpose. Poly(dimethyldiallyl ammonium chloride) and its copolymers are also sometimes used. The cationic polymers form water-insoluble complexes with anionic surfactants, but the complexes are solubilized again by the excess amount of surfactant in the shampoo formulations (Goddard, 1990). The complexes precipitate out of the shampoo solution when diluted by washing and/or rinsing, and then they attach to the hair surface. The attached complexes work as a lubricant for the hair fibers. Figure 4.15 illustrates schematically the solution behaviors of the complex. Although the smooth and lubricant feeling of the hair is normally obtained with an after-shampoo hair conditioner, both cleansing and conditioning can be obtained with the so called 2-in-

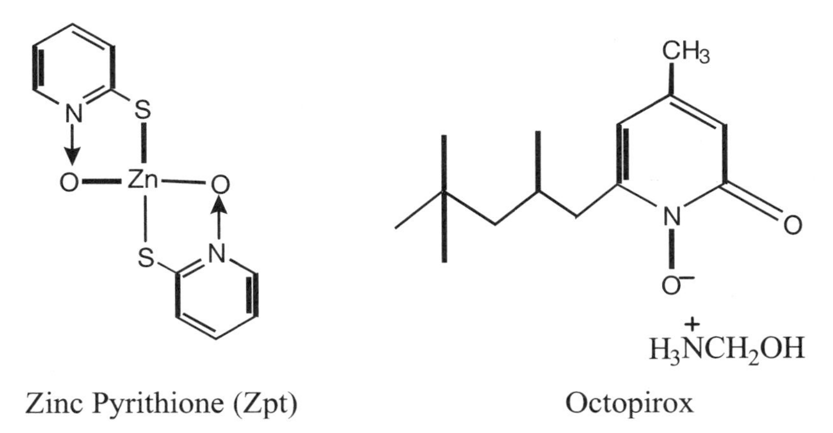

Figure 4.14 Chemical structures of typical antidandruff agents used in shampoo formulations.

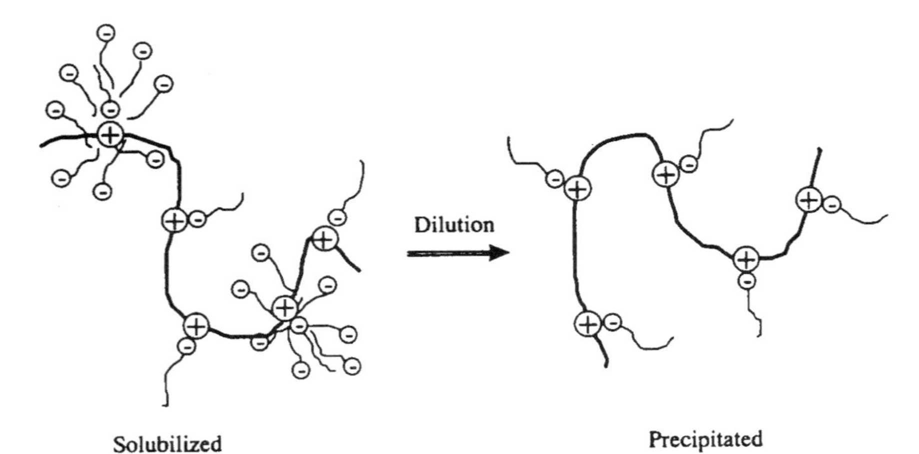

Figure 4.15 Schematic representation of the mechanism of conditioning effect of shampoos. The solubilized complex between the cationic polymer and anionic surfactant precipitates on hair fibers when the shampoo is diluted by washing and/or rinsing.

1 shampoos. In such shampoos, dimethylpolysiloxane (silicone) oils are frequently blended to give hair the conditioned feeling.

b. Toiletry Soaps

Soap is the oldest cleansing agent, and even used to be a main component of heavy-duty detergent. Today soaps are used mostly as toiletry soap bars. Soap bars are almost pure sodium salt of fatty acids containing about 10% water and a small amount of perfumes and pigments. Some bactericides are included in deodorant soaps. Fatty acids from tallow and coconut oil can be blended in an appropriate ratio to control the solubility (the Krafft point) of soaps. Palm oil fatty acids have also been used recently. Combination bars are a mixture of soap and synthetic surfactants that have a high-foaming property and give different feelings during washing and rinsing. Alkyl or alkylpolyoxyethylene sulfates and acyl isethionates are commonly mixed with soap.

c. Face Cleansers and Body Shampoos

This new category of liquid cleansers has an elegant image and is convenient to use. Young women are the major consumers for this category of products. Mild surfactants that do not irritate the skin, such as monoalkyl phosphate, are particularly selected for these cleansers.

4.2.4 LIGHT-DUTY DETERGENTS

Light-duty detergents include household cleansers for dishes, kitchenware, baths, toilets, and glass, as well as general-purpose detergents for living rooms and so on. Because the target soils or dirt vary considerably, a wide variety of cleansers are needed; no general principle to clean them all is known. Following is an explanation for each detergent.

a. Dishwashing Detergents

These detergents can be used to wash dishes, vegetables, and fruits. Because the soils to be washed are limited in both volume and variety (compared with the soils on clothing), the formulations of dishwashing detergent are relatively simple: They are basically just aqueous solutions of a surfactant in a neutral pH. The main surfactants are anionic alkylpolyoxyethylene sulfates, linear alkylbenzene sulfonates, alkyl sulfates, α-olefin sulfonates, nonionic polyoxyethylene alkyl ether, and alkyl polyglucoside. Some cosurfactants are used as boosters to enhance the detergency and foaminess of anionic surfactants. Alkyldimethylamine oxide, alkylcarboxy betaine, and alkanolamide of fatty acids are frequently used for this boosting purpose.

In the selection of the surfactants for dishwashing detergents, a special goal is low skin irritation by the surfactant. For example, alkyldimethylamine oxide, used as a booster for detergency and foaminess, is known to depress the skin irritation factor of anionic surfactants. These synergistic effects of this amphoteric surfactant with an anionic one are probably due to the complex (addition compound) formation between them (Kolp *et al.*, 1963; Rosen *et al.*, 1964). Research is still being done to find new surfactants that cause *no* irritation to hand skin but have high detergency and foaminess.

b. Household General-Purpose Detergents

Aqueous surfactant and alkali solutions are utilized as a detergent for walls, floors, furniture, and so on. They are used after dilution. Stronger detergents are available for heavy dirt in gas ranges, kitchen fans, etc. In these detergents, organic solvents such as cellosolves and carbitols (one or two moles of ethylene oxide adducts of short chain alcohols) are additionally blended. Because the surfactants used in this type of detergent must be stable in alkaline condition, linear alkylbenzene sulfonates and polyoxyethylene alkyl ether type agents are normally selected.

c. Bath Cleansers

Bath cleanser contains a special component—a chelating agent for multivalent metal ions—since the main dirt on bathtubs and pails consists of calcium or magnesium salt of fatty acids (metal soaps or scum). Chelating agents such as EDTA, citric acid, and malic acid are quite able to decompose the metal soaps and make them soluble in water. Typical surfactants such as alkylbenzene sulfonates, alkyl sulfates, alkylpolyoxyethylene sulfates, and the polyoxyethylene type nonionic surfactant are formulated in this detergent together with some organic solvents.

d. Glass Cleansers

Glass cleanser is basically a diluted solution of the general-purpose detergent consisting essentially of surfactant and alkali. Special attention is paid to prevent dripping of the foams attached to the glass surface and streaking on the cleaned glass.

e. Toilet Cleansers

There are three types of toilet cleansers: acid type, alkaline type, and neutral type containing dispersed inorganic powders. The acid-type cleanser incorporates hydrochloric or sulfuric acid and decomposes inorganic substances, such as calcium phosphate in urine grime. The alkaline-type cleanser contains hypochlorite and sodium hydroxide and decomposes the binder components of organic dirt. It is quite dangerous to mix acid- and alkaline-type cleansers; the chemical reaction that occurs generates deadly chlorine gas. A chamber pot is cleaned by polishing it with the inorganic powders in the neutral-type cleanser. A small amount of surfactant (1–2 wt %) is blended in all types of cleanser to enhance wetting and dispersion of dirt.

4.2.5 DETERGENTS FOR PROFESSIONAL AND INDUSTRIAL USE

In this section we will discuss a few of the many uses of detergents in the professional and industrial fields. Washing or cleaning is done in every professional and industrial activity. In almost all such cleansing processes, surfactants, of course, are used. Thus, better applications and selections of the agents for each application field are always under investigation.

a. Dry Cleaning

Washing with organic solvents instead of water is called *dry cleaning.* Oily dirt, which is hard to clean up in aqueous medium, is easily removed in organic solvents by dissolving them. On the other hand, water-soluble soils such as inorganic compounds and proteins are difficult to remove with the dry cleaning method. Thus, surfactants are used to accelerate the removal of water-soluble soils. The surfactants used for dry cleaning must be soluble in organic solvents and have a hydrophobic nature (low HLB number). The surfactants solubilize the water in the organic solvents, and the solubilized water adsorbs the hydrophilic soils and substrates to separate them easily. Secondary solubilization of water-soluble soils into the solubilized water is also supposed to be useful for detergency, but this process may add only a small contribution. Dry cleaning is mainly applied to wool and silk textiles since these shrink if washed in an aqueous alkaline medium. Leathers, furs, and fashion wear are also washed by the dry cleaning method.

The surfactants used in the dry cleaning process are soaps (ammonium or ethanolammonium salts for an oil-soluble property), petroleum sulfonates, linear alkylbenzene sulfonates, and dialkyl-sulfo succinates as anionic surfactants, and polyoxyethylene alkyl ethers, polyoxyethylene nonylphenyl ether, and polyoxyethylene sorbitan fatty acid esters as nonionic ones. Cationic surfactants are sometimes added to obtain antistatic and softening effects. The organic solvents are tetrachloroethylene, trichloroethane, trichloro-trifluoroethane, and petroleum (saturated hydrocarbons). The halogen-type solvents will be prohibited or at least limited for use in the near future due to the environmental problems.

b. Cleansing in the Electronics Industry

An integrated circuit is now so high in density that it is constructed with devices of submicrometer size. It can be easily imagined that the presence of dusts and stains on silicon wafer can cause fatal defects to these circuits. Therefore, all the processes of LSI production are performed in a clean room, and the water used is an ultra-pure one.

A number of complicated manufacturing processes are necessary in LSI production: coating and etching of photoresists, formation and etching of ultra-thin oxide films, ion injections, etc. Since each manufacturing step is a possible origin of pollution, a cleansing treatment is necessary after each step. An oxidative acid or alkali solution, organic solvents, and ultra-pure water (for rinsing) are mainly used. The oxidative acid solution is a mixture of hydrogen peroxide and sulfuric or hydrochloric acid. The oxidative alkaline solution contains ammonia and hydrogen peroxide. Trichloroethylene, tetrachloroethylene, isopropanol, and ethanol are the most popular organic solvents used. Organic solvents containing chlorine atoms, however, will be severely limited or prohibited in the near future because of the environmental problem of ozone layer destruction.

Water-based detergents for electronics cleansing are now under development. One new type, polyoxyethylene alkyl ether with a relatively short alkyl chain, is used as an organic solvent to dissolve stains. The nonionic surfactant phase containing the organic stains can be readily separated from the water phase after rinsing on heating, utilizing the cloud point of the agent in aqueous solutions (see Section 3.3.2.*b*). The separated water then can be recycled for rinsing.

c. Detergents in the Pulp and Paper Industry

The cleansing processes in this industry are mainly deresination, felt washing, and deinking. Deresination means to remove resins such as resin acids, fatty acids, dyes, tannin, hemicellulose, etc., to prepare the dissolving pulp (which is raw material for rayon, cellophane, and carboxymethyl cellulose (cmc)). The dissolving pulp must be a higher quality than that for papermaking, thus a deresination process is necessary. The surfactants for this purpose must be stable in acidic or alkaline conditions at high temperature (130–150°C) and pressure. Polyoxyethylene alkyl ether or polyoxyethylene nonylphenyl ether is selected to satisfy the required conditions.

Felt is used in the paper industry as a kind of large blanket on which wet paper is carried just after a paper sheet is formed from dispersed pulp. Many kinds of soils and stains present in the papermaking processes attach to the felt: Pitch and tar materials from the pulp, sizing agents, fillers, and fixers are examples. Polyoxyethylene-type nonionic surfactants, linear alkylbenzene sulfonates, and their mixtures are used as felt cleansers.

A deinking agent works to detach ink from used papers—newspapers in particular—and to renew the pulp. Deinking is a kind of cleansing, but we will discuss it in Section 4.3.1.*c* since the adsorption of the surfactant is essentially important in this process.

d. Scouring and Soaping in Textile Industry

There are two kinds of cleansing processes in textile industry: scouring and soaping. Scouring is a prewashing of fibers to remove waxes, fats, proteins, and soils from natural fibers such as cotton, wool, and silk, and to remove spinning oils, sizing pastes, and weaving and knitting oils from synthetic textiles. This prewashing (scouring) of fibers is followed by treatments of bleaching and dyeing. Soaping, on the other hand, is done after dyeing and textile printing to remove unfixed dyes, auxiliary chemicals, and printing pastes.

Scouring of cotton is carried out in an alkaline solution containing 0.1–0.3 wt % surfactant at about 120°C for several hours. The surfactants used in this process must be chemically stable in alkaline solutions, and polyoxyethylene alkyl ether, sulfonate-type anionic surfactants, etc., are selected. The cloud point of nonionic surfactants must be high enough to dissolve at high temperatures. Enzymes such as amylase are sometimes utilized to decompose paste stains. Sodium carbonate and surfactants are used for scouring of wool; polyoxyethylene alkyl or alkylphenyl ether, alkyl sulfate, and their mixtures are the agents for this purpose. Soap is the common agent for scouring of silk, with nonionic surfactants sometimes used as cosurfactants. Sericin is the main dirt of silk, so enzymes are sometimes blended with a nonionic surfactant for the silk scouring process. Nonionic surfactants are mainly used for the scouring of synthetic fibers because of their strong detergency for oils.

Soaping can be done with anionic, nonionic, and cationic surfactants, depending on the kinds of textiles, dyes, and soaping conditions. Anionic surfactants such as soaps, alkyl sulfates, and naphthalene sulfonate formaldehyde condensate are selected for acrylic knitted goods. A nonionic surfactant such as polyoxyethylene alkylamine is used for nylon fabrics. Cationic surfactants are also sometimes used for cationic dyes. A blended agent of anionic/nonionic surfactant is useful for cotton fabrics. The proper soaping agent is still selected by the method of trial and error; no guiding principle or theory has yet been found.

e. Metal Cleaning

The cleaning of metal surfaces is necessary for subsequent processes such as annealing, painting, and galvanization. Rusty stains, oils, and grease must be removed from the metal surface. Alkaline cleaning is most popular, and the cleaning agent consists of alkali and 1–5 wt % surfactant. Alkalis such as sodium hydroxide, sodium carbonate, sodium silicate, and sodium phosphate are commonly used. The surfactants used are polyoxyethylene-type nonionic and sulfonate-type anionic ones, both of which are stable in alkaline solutions. Acid cleaning to remove the rust is done using hydrochloric acid, sulfuric acid, phosphoric acid, etc.

Emulsion cleaning is especially useful to remove oily dirt. Organic solvents—such as kerosene and naphtha—and emulsifying agents (surfactants) are mixed and used. Steel plates are dipped in the mixed solvent and rinsed with water. Emulsion is formed in this rinsing process. Premixed emulsion is also used as a cleaning agent. The surfactants employed for this purpose are ethanolammonium oleate, alkylbenzene sulfonates, alkyl sulfates, polyoxyethylene alkyl ether, sorbitan fatty acid esters, and so on. The advantage of emulsion cleaning is the low reactivity of the agents with aluminum, zinc, etc. (compared with alkaline cleaning).

f. Cleansing in Miscellaneous Fields

Surfactants are always used in any cleansing process and so are used in a variety of fields not mentioned previously. Examples are automatic dishwashing systems (in large-scaled restraint), building maintenance, car washing systems, medical instruments, membrane separation processes, and so on. Surfactants possess many functions required for detergency and always will be employed even in the future when cleansing is needed.

4.3 Applications Utilizing Adsorption Phenomena

The adsorption phenomena of surfactants are applied in many areas of daily life as well as industrial processes. They are used in more surfactant applications than are detergents, although detergents generate higher sales. The functions and the related application examples of surfactant adsorption are listed in Table 4.3.

4.3.1 APPLICATIONS OF GAS/LIQUID INTERFACE ADSORPTION

Surface tension lowering and foaming/defoaming are two typical phenomena caused by surfactant adsorption at gas (air)/liquid interfaces. Surface tension lowering gives remarkable effects on the wettability of liquids, but this phenomenon will be discussed in Section 4.3.5 since the interfacial tensions at liquid/solid and liquid/liquid also affect it. In this section we will focus only on applications of the foaming and defoaming phenomena.

Table 4.3 Adsorption Phenomena of Surfactant and Their Functions and Applications

Interface	*Function*	*Application Examples*
Gas/liquid interface	Surface tension lowering (wetting)	Spreader for agricultural chemicals; Antifogging agent for greenhouse, window glass, etc.
	Foaming and defoaming	Shampoo; Detergent; Fire-extinguishing agent; Deinking agent; Mineral flotation; Shortening; Defoaming agent
Liquid/liquid interface	Interfacial tension lowering (wetting)	Detergent (rolling-up of oily dirt); Fire-extinguishing agent
	Emulsification	Detergent (dishwashing); Shampoo; Cosmetics (skin cream, lotion); Butter; Margarine; Mayonnaise; Agricultural chemicals (emulsion type); Emulsion and suspension polymerization; Asphalt; Tertiary oil recovery
Solid/liquid interface	Wetting	Detergent; Water- and oil-repellent treatment; Mineral flotation; Rust-preventive agent; Spreader for agricultural chemicals; Antifogging agent; Dyeing auxiliaries for fabrics
	Dispersion	Ink; Paint; Cement dispersant; Detergent; Magnetic recording materials; Magnetic fluid; Pigment dispersion for paper coating; Agricultural chemicals (flowable type); Antiscaling agent for desalination; Coffee creamer
	Coagulation	Water treatment in water-supply and drain systems; Fixers in papermaking process; Coal-oil mixture
	Lubrication	Metal rolling oils; Hair conditioner
Solid/gas interface	Wetting	Water- and oil-repellent treatment
	Lubrication	Fabric softener; Hair conditioner
	Antistatic	Antistatic agents for plastics and fabrics

a. Foaming/Defoaming Agents and Fire Extinguishing by Foams

Several commercial products require high foaminess—soaps, shampoos, body shampoos, dishwashing detergents, etc. These are all in the category of detergents and were discussed in the previous section. The mechanisms of foam stabilization by the adsorption of surfactants and polymers have also already been discussed (Section 3.2.2.*a*).

In most industrial processes and/or professional operations, foaming is generally not preferred. An exception is fire extinguishing by foams. Fire-extinguishing agents are mainly used for fires of organic solvents such as petroleum. The foam covers the surface of the burning solvent and shuts out the air. Using aqueous foams to cool the burning material is also effective in firefighting. Firefighting foams must be strong enough to persist even in contact with the fire and the organic solvents. Protein derivatives like a hydrolyzate of keratin and synthetic surfactants are utilized as foaming agents. The protein derivatives are denatured by the fire at the surface of the mass of foams, which makes the foams solid. This solid foam is stable in the fire, but does not spread rapidly on oil surfaces. Synthetic surfactants such as alkyl sulfates and alkylpolyoxyethylene sulfates show a high-foaming property and are mostly used for fighting building fires because their foams are relatively weak in organic solvents.

Fluorinated surfactants are often formulated as a cosurfactant in fire-extinguishing agents. As already pointed out in Section 4.1.3, some mixtures of fluorocarbon- and hydrocarbon-type surfactants show phase separation into two kinds of micelles. This phase separation phenomenon works favorably in firefighting foams (Shinoda and Nomura, 1980). Figure 4.16 illustrates schematically the mechanism of fire fighting by the mixed surfactant systems of fluorocarbon and hydrocarbon. One phase, rich in the fluorocarbon surfactant, adsorbs at the air/water interface, and the other, rich in the hydrocarbon surfactant, does so at the water/oil interface. As a result, both interfacial tensions of γ_W and γ_{OW} become low and the water layer (foam) spreads rapidly on the oil surface. Of course, water alone does not spread on the oil phase, which has a lower surface tension than water. Thus, the phase separation phenomenon in the fluorocarbon/hydrocarbon surfactant mixtures works to enhance the wetting and spreading of water on the oil phase.

Defoaming is another important application of adsorption at air/water interfaces. Foams are a nuisance in most industries, so defoaming agents are frequently employed in the industrial fields of pulp and paper, fermentation, rubber and plastics, fabrics, and so on. The defoaming mechanism and some defoaming agents were discussed in Section 3.2.2.*b*.

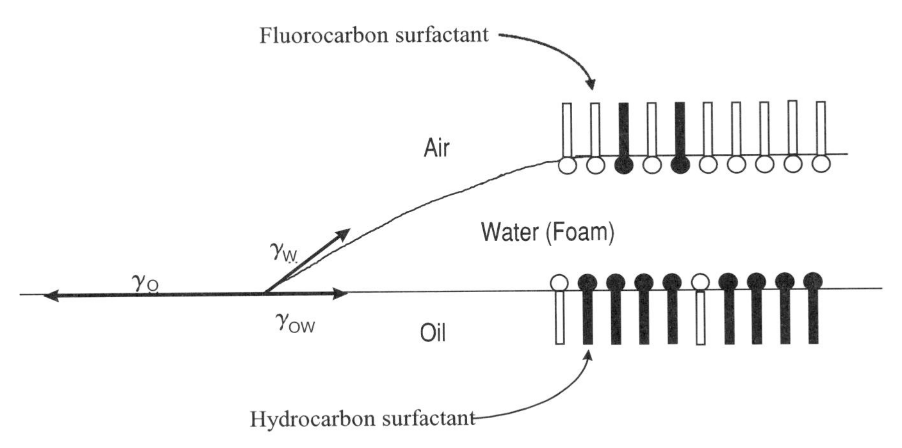

Figure 4.16 Surfactant mixtures of fluorocarbon/hydrocarbon are effective as fire-extinguishing agents. Both γ_W and γ_{OW} are depressed by the adsorption of the two kinds of surfactant.

b. Mineral Flotation

Raw ore is a mixture of metallic minerals and useless ones with no metal element. Flotation, a clever method for collecting only the valuable metallic minerals from the ore mixtures, uses a technique of surface activity. Ore particles are ground into fine powders several tens of micrometers in size. When the fine powders are dispersed in water, only the metallic minerals are hydrophobized by the adsorption of some surface active materials. The hydrophobized powder particles then attach themselves to air bubbles in water due to their hydrophobic surfaces. These mineral-bearing bubbles then come up to and float on the water surface and are collected. Figure 4.17 shows a schematic representation of the mineral flotation process. The surface active hydrophobizing materials that must attach selectively to the useful metallic minerals are called *collectors*. The collector molecules adsorb on the surface of the metallic mineral particles and orient their hydrophobic chain to the water medium.

The collectors are alkyl dithiocarbonates (ROC(=S)SM: xanthate) and dialkyl dithiophosphates ($(RO)_2P(=S)SM$) for sulfide minerals. They adsorb on the minerals by chemical affinity (chemisorption). Soaps, alkylamine acetate, alkyl phosphates, alkyl sulfates, etc., are employed for nonsulfide minerals. Some of these collectors are typical surface active agents and work also as foaming agents. Pine oils, medium-chain alcohols, etc., are sometimes used to generate

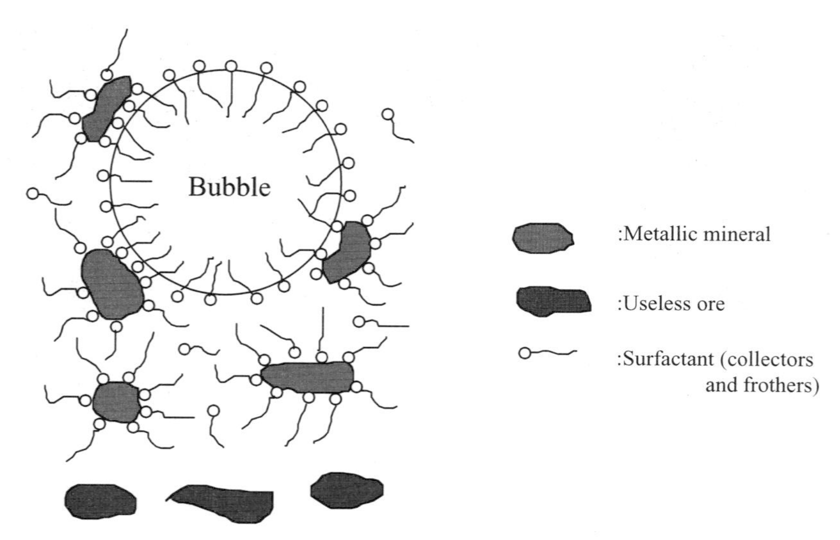

Figure 4.17 Schematic illustration of a mineral flotation process.

small air bubbles; these agents are called *frothers*. Regulators—such as acid and alkali (to control pH value), multivalent metal salts, inorganic salts, water-soluble polymers, etc.—are often used to give a selective adsorption property to the collectors. The regulating mechanism of the selective adsorption is not clear at present.

c. Deinking Agents in Paper Recycling

Waste or used papers (newspapers in particular) are recovered again to pulp in considerable amounts. Recovery and recycling of used papers is, of course, important for protection of wood resources and forests and will become ever more so in the future. Deinking is a key technology for the recycling of waste newspapers. This deinking technology is quite similar to that of mineral flotation.

A photograph of a deinking machine is shown in Figure 4.18. Hydrophobized ink particles are attached to air bubbles in a water medium and swept away together with the foams at the top of a deinking tank. Surfactants called deinking agents are used to remove the ink from the redispersed pulp of the waste papers, to give the hydrophobic nature to the ink surfaces, and to generate foams. Fatty acids are employed as a hydrophobizing agent and polyoxyethylene-type nonionic surfactants are used for foaming. Adducts of both polyoxyethylene and poly-

Figure 4.18 A deinking machine. Hydrophobized ink particles attach themselves to air bubbles, come to the top of the solution, and are swept away.

oxypropylene to triglyceride/glycerin mixtures or fatty alcohols are also frequently utilized as deinking agents. In these cases, it is not easy to distinguish between the hydrophobizing and foaming agents.

4.3.2 APPLICATIONS OF LIQUID/LIQUID INTERFACE ADSORPTION

Emulsion is the main (almost only) application of surfactant adsorption at liquid/liquid interfaces. The basics of emulsion were explained in Section 3.2.3, and here we focus only on the applications of the emulsions. Emulsions are used in a number of commercial products that contain some oils for various aims. Following are typical examples of these emulsion products.

a. Food Emulsions

There exist many emulsions in foods. Typical examples are butter and margarine as a W/O emulsion, and milk, mayonnaise, and dressing as an O/W one. Ice cream is also a frozen O/W emulsion containing air bubbles of about 50 volume

%. Triglycerides such as soybean oil, corn oil, etc., are commonly used as oil components in the food emulsions. The surfactants (emulsifiers) used in food emulsions must be food additives. Examples are monoglyceride, polyglycerin fatty acid esters, sugar ester of fatty acids, sorbitan fatty acid esters (SPAN), polyoxyethylenesorbitan fatty acid esters (TWEEN), propyleneglycol fatty acid ester, lecithin, and mono- or diglyceride esters of lactic, tartaric, or citric acid. TWEEN is a food additive in the United States and Europe that is not permitted in Japan. In natural emulsions such as milk, proteins such as casein often work as an emulsifier. In butters and margarine, the continuous oil phase is a kind of gel consisting of thin crystalline networks of solid fats and liquid oils among them. Thus, the water droplets in the emulsions cannot move around, and the emulsion is very stable as a result.

Oils and fats provide savory taste to foods and the consumption of oils and fats is increasing in rich countries. Much consumption of oils and fats, however, results in the health problem of obesity. Emulsion technology may be able to contribute to the solution of this problem. For example, a tasty margarine with low fat content could be a research target of emulsion scientists.

b. Cosmetics and Toiletries

Skin-care cream, skin milk, foundations, 2-in-1 shampoos, and hair conditioners are examples of emulsions in the cosmetics and toiletries industries. For these, combinations of SPAN and TWEEN are frequently employed for emulsifiers. Polyoxyethylene alkyl ethers are also often used. Liquid crystal emulsification is a unique technology that makes very stable emulsions (Suzuki *et al.*, 1984; Suzuki and Tsutsumi, 1987; Suzuki *et al.*, 1989) and will be discussed in Section 4.4.2.*a*.

Polydimethylsiloxane (silicone) oils and fluorocarbon oils have recently become popular in cosmetic and toiletry products. Typical examples of these oils are shown in Figure 4.19. Silicone oils are water-repellent and give a unique slippery feeling with low viscosity. These properties are utilized for sweat-resistant foundations, 2-in-1 shampoos, and so on. Fluorocarbon oils, which are more water-repellent, are used in some foundations and lipsticks. To emulsify the silicone oils, a surfactant (emulsifier) having a silicone chain as a hydrophobic group is, of course, more powerful. The hydrophobic group of silicone chain has more affinity for the silicone oils due to their similar solubility parameters and consequently adsorb more effectively. Silicone-type surfactants are useful for removing silicone-oil foundations from the face on cleansing. These situations are the same in the fluorocarbon-type oils and surfactants. An interesting problem yet to

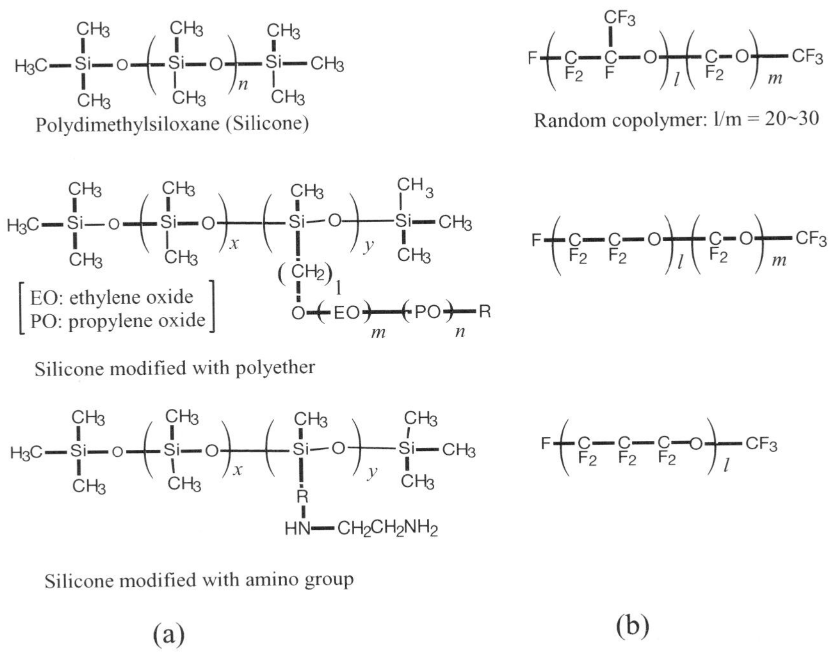

Polydimethylsiloxane (Silicone)

Random copolymer: l/m = 20~30

EO: ethylene oxide
PO: propylene oxide

Silicone modified with polyether

Silicone modified with amino group

(a)

(b)

Figure 4.19 Chemical structures of typical polydimethylsiloxane (silicone) (a) and fluorocarbon (b) oils used in the field of cosmetics and toiletries.

be solved is whether the HLB concept holds in the emulsification of such oils by the silicone- and/or fluorocarbon-type surfactants.

c. Agricultural Chemicals

Most agricultural chemicals—germicides, insecticides, herbicides, etc.—are water-insoluble organic materials. However, they cannot be applied to plants as a solution of organic solvents because such solvents are harmful to plant leaves. So organic solutions of agricultural chemicals must be emulsified in water and applied as O/W emulsions. The commercially available products are commonly supplied as solutions of agricultural chemicals and surfactants in some appropriate organic solvent. Users then can easily emulsify the organic solutions by just diluting with water. Hydrocarbons such as toluene, xylene, kerosene, alcohols like ethanol, isopropanol, and ketones are employed as organic solvents to dissolve the agricultural chemicals. Blended combinations of nonionic surfactants like polyoxyethylene alkyl ethers and anionic ones like a calcium linear alkylbenzene sulfonate are usually used for the emulsifier.

d. Suspension Polymerization

When water-insoluble monomers such as styrene and acrylic acid esters are polymerized as emulsion droplets containing an initiator, polymer particles of spherical shape are obtained. This polymerization technique is called *suspension polymerization.* Aqueous solutions of water-soluble monomers such as acrylic acid and acrylamide are emulsified in an organic solvent and polymerized; this is called *inverse suspension polymerization.* In suspension polymerization, the monomers are polymerized in the emulsion state, but we do not call it emulsion polymerization. Emulsion polymerization is a different process (which is discussed in Section 4.4.1.*a*) in which the solubilization of monomers into surfactant micelles is essentially important. The terminology is a little complicated; take care that you are not confused!

The main advantages of suspension polymerization are that the heat of polymerization can be easily controlled compared with bulk polymerization, that powdered polymers are readily obtained, and that the spherical polymer particles can be synthesized. Figure 4.20 shows an electron microscope photograph of polystyrene spherical particles beautifully monodispersed. Such monodispersed spherical particles can be applied, for example, to a spacer of liquid crystalline display, where the liquid crystal is sandwiched between two transparent glass electrodes. The two glass plates must be parallel everywhere in the display panel with a distance of 5–10 μm between them. The spacer made of monodispersed polymer particles is used to maintain the homogeneous thin distance between the two electrodes.

Inverse suspension polymerization is particularly powerful in manufacturing super-water-absorbing polymers. The polymer absorbs water as much as several hundreds to one thousand times its own weight, forming a hydrogel. These super-water-absorbing polymers are used for disposable diapers, sanitary napkins, and the like. Cross-linked sodium polyacrylate, a typical example of a super-absorbing polymer, is synthesized by the inverse suspension polymerization of an aqueous solution of sodium acrylate and a cross-linker. If we were to employ the solution polymerization method for the production of these polymers, the entire solution would have to be gelled in the whole reaction vessel and the product would be unwieldy. This is one of the best examples to support using the inverse suspension polymerization instead.

There is no hard-and-fast rule for selecting the proper emulsifiers for the sus-

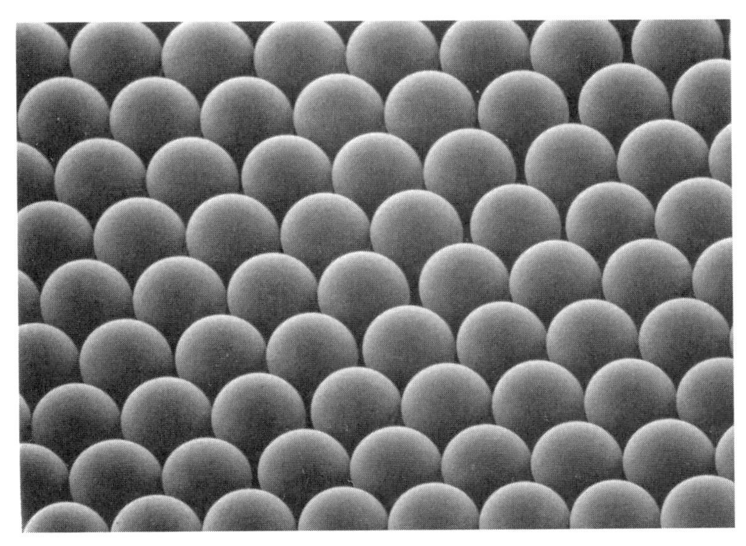

Figure 4.20 An electron microscope photograph of monodispersed polystyrene particles synthesized by a suspension polymerization. The diameter of these particles is about 6 μm.

pension and the inverse suspension polymerization methods. Common surfactants are, of course, also useful for this purpose; but sometimes must be avoided. In the applications of liquid crystalline display, for instance, ionic substances are prohibited because they cause an electric current between the two electrodes. So ionic surfactants cannot be used in the synthetic process of monodispersed polymer particles. An inorganic powder such as a calcium phosphate is sometimes employed as a stabilizer for monomer emulsions. The inorganic powder particles adsorb on the surface of the monomer droplets and keep the droplets from coalescencing. Polymeric materials such as cellulose derivatives are often used to stabilize the emulsions in the inverse suspension polymerization process.

e. Other Applications of Emulsions.

Asphalt is applied to pave a street in two ways. One is a hot-melt method, and the other is an emulsification one. In the emulsification method, cationic surfactants like fatty amines are employed as emulsifiers. Silica and calcium carbonate are used as aggregates together with asphalt in the pavement. Their surfaces are negatively charged in aqueous systems, and the asphalt emulsion stabilized by the cationic surfactants is destroyed on contact with the aggregates. This coagulation takes place on the road surface in the paving process.

In metal working, cutting and rolling oils lubricate and cool the metal surfaces. The O/W emulsions of mineral oils, tallow, and palm oils are used for this purpose. Oils give the lubrication effect, and water gives the cooling effect. The emulsifiers used are ammonium salt of fatty acids, nonionic surfactants, and so on.

Tertiary oil recovery is also an application of emulsion technology. The emulsifiers for this process must be extremely inexpensive, and oils whose aromatic components are sulfonated are frequently employed.

4.3.3 APPLICATIONS OF SOLID/LIQUID INTERFACE ADSORPTION

a. Applications of Dispersing Action

Dispersion is a homogeneous mixture of solid fine particles and liquid (see Section 3.2.4). The dispersed particles are stabilized by electrostatic repulsion and/or steric repulsive forces of adsorbed polymers (Sections 3.2.4.*b* and 3.2.4.*c*). The theories of dispersion stability are always constructed on the given assumption that electric charges or adsorbed polymers preexist at the surface of the dispersed particles. A very important but difficult problem, however, in developing the dispersing agents is how to put the charges and/or the polymers onto the particle surfaces. A key point is how to make the dispersing agent adsorb onto the particles. The origin of the dispersing action of the agents is their solubility into medium liquids. Either electrostatic or steric repulsion results from the osmotic pressure due to the higher concentrations of counterions or polymer chains between the two approaching particles. This mechanism works, of course, only when the ionic groups or the adsorbed polymers are miscible with solvents to form solutions. The entropic repulsion of adsorbed polymers results from the conformational change of polymer chains and is also effective only in the solutions. The dispersing agents must be soluble in the liquid medium but also must adsorb on the particle surface without dissolving too much into the solvent. This is the challenge in the development of the dispersants. Metathetic reaction, electrostatic attractive force, hydrophobic interaction, and so forth are frequently used to enhance the adsorption of the agents onto the surface of the particles to be dispersed.

Dispersion of pigments. There are many kinds of commercial products in which pigments are dispersed—ink, paint, cosmetic foundation, lipstick, paper coatings, (which use inorganic pigments) and so on. Which dispersing agent is suitable for each pigment of course depends on the organic or inorganic nature of the pigment. Organic pigments contain the aromatic moiety as a chromophore in their molecule, and the dispersant fitted for them also contains the aromatic groups. The aromatic portion of the agent has an affinity for that of the pigment molecules, which enhances the adsorption of the dispersing agent on the pigment surface. Oligomer-type dispersants such as formaldehyde condensates of naphthalene sulfonate and the condensates of alkyl naphthalene sulfonate are typical. For disperse dyes, in particular, phenol sulfonic acid salts, formaldehyde condensates of naphthalene sulfonate, lignin sulfonate, etc. are employed. The dispersing agents commonly used in nonaqueous solvents for paints and inks are alkylbenzene sulfonates and alkylnaphthalene sulfonates. Divalent metal ions like those of calcium and magnesium are selected as the counterions for the above dispersants to enhance their solubility into organic solvents.

Kaolin, clay, calcium carbonate, aluminum hydroxide, titanium dioxide, and ferric oxide are typical inorganic pigments. The main dispersing agents for these inorganic pigments are sodium polyacrylate, copolymers of olefin and sodium maleate, and polyphosphates. The carboxyl or phosphate groups in these dispersants show the affinity for metal ions, which causes their adsorption on the inorganic pigments. Sodium polyacrylate, for instance, is assumed to adsorb on calcium carbonate by an ion-exchange reaction to form calcium salt of the carboxyl group. Dispersions of inorganic pigments are most popularly used in the application of paper coatings. Art paper used for printed materials such as posters and calendars is a typical coated paper. White pigment like kaolin or calcium carbonate is coated on the art papers together with a binder polymer. The readers may have noted at some point that when art papers are burnt, the smoke and flames are similar to those of plastics and a considerable amount of ash remains. This is because the binder polymer burns like a plastic and the ash is the inorganic pigment. To manufacture art papers having a smooth and bright surface, the pigments must be dispersed well, requiring a dispersing agent of high performance.

Cement dispersant. One of the biggest markets for dispersing agents is concrete manufacturing. Concrete slurry consists of cement, sand, gravel, and water. Handling of the slurry is impossible when the amount of water is too small, and the final concrete is too weak when the amount of water is too large. However, a cement dispersing agent can fluidify the concrete slurry even when too little water is used and strengthen the final concrete. The most popular cement dispersant is formaldehyde condensates of naphthalene sulfonate. An agent having higher molecular weight than that of the pigment dispersant is employed for this purpose; molecules with about 10 naphthalene moieties show the best performance. Maleic acid/α-olefin copolymer, lignin sulfonate, polyacrylic acid partially modified with polyoxyethylene chain, and the like are also used. Cement dispersants have been used mainly in manufacturing poles, cross-ties, pipes, and so on, since they are produced under controlled industrial conditions and processes. However, recently developed dispersing agents expand their application to the outdoor construction of buildings, bridges, etc. One such agent, a high-polymer type, provides "super-flowing concrete," which flows well and fills up even small spaces of buildings without any mechanical vibrations. Another technology is a kind of controlled release of the dispersant that extends the handling time of concrete after the mixing of components. With buildings becoming taller and taller, higher-performance cement dispersant will be required in the future.

Magnetic recording materials. Magnetic recording films, such as floppy disks and magnetic tapes, consist of a base film of polyester (polyethylene telephthalate: PET) and a magnetic layer. Figure 4.21 illustrates schematically the structure of a magnetic film. Magnetic powders are ultra-fine particles of ferric oxide or metal iron with a diameter less than 0.1 μm that bear the magnetic recording. For high-density recording, the magnetic powders must be well dispersed and present as a primary particle without aggregation in the magnetic layer. If the size of the aggregate is larger than that of 1 bit recording, an error will occur at the position of the aggregate. Thus, the good dispersion of the magnetic particles in a coating mixture of magnetic powder, binder, and organic solvents is a key point in the manufacturing of high-density recording magnetic films. The magnetic powders are ultra-fine particles with a magnetic dipole and thus are hard to

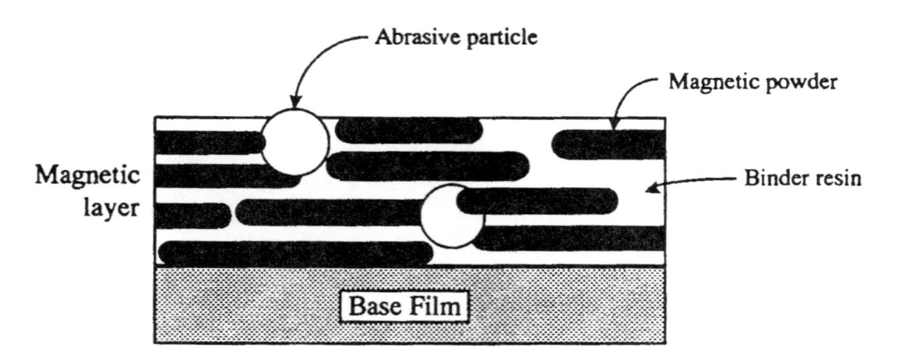

Figure 4.21 A schematic illustration of a magnetic tape.

disperse because of the strong attractive forces of van der Waals and magnetic interactions between particles. Currently there are two methods for dispersing the magnetic powders. One is a surface treatment of magnetic powders with hydrophobizing agents such as silane coupling agents. The other is the introduction of some polar or ionic groups into the binder polymers: Sulfonic acid and carboxyl groups are frequently attached to the molecules of polyurethane and polyvinylchloride. Both methods reduce the high interfacial energy between the magnetic powders and organic solvents.

One can understand from the above explanation that evaluation of the degree of dispersion of the magnetic powders is very important. But unfortunately, almost no method is available for detecting the size distribution of magnetic powders both in the coating mixtures of liquid state and in the coated solid magnetic layers. An optical method such as light scattering is useless because of the black color of the magnetic dispersions. The Coulter counter method cannot be applied directly to the concentrated dispersions of magnetic powders, and the dilution of the dispersions may change the size distribution of the system. In addition, the electric conductivity is too small for the Coulter method since the magnetic powders are dispersed in organic solvents. The small-angle X-ray scattering technique is a possible method, but no trial has been reported so far. Consequently, the practical evaluation of coated magnetic films—such as the output intensity of recorded signals, hysteresis loop in the magnetization-magnetic field curves, surface roughness, etc.—is currently the only reliable method for checking the dispersion of the magnetic powders in the films.

The rheological properties of magnetic paints govern the coating process of the magnetic films and depend highly on the dispersion of the magnetic powders. The degree of dispersion of the magnetic powders affects the viscoelastic nature of the paint and the coating behavior as a result. The most suitable viscoelastic property depends on the coating methods such as gravure, offset, and die. There is no guiding principle for preparing the best dispersion of magnetic powders for each coating method; the best coating conditions are chosen by trial and error in the practical production. A few microscopic theories have been given on the rheological properties of the magnetic suspensions. One of them deals with the magnetic suspension as a network structure model of flocs of magnetic powders and interprets the rheological properties by this model (Kanai and Amari, 1993).

Magnetic fluid. Magnetic fluid is a condensed suspension of magnetic powders with a diameter of about 10 nm in an organic solvent. The fluid is regarded as a ferromagnetic liquid and is attracted by a magnet. The real substance that is attracted by the magnet is, of course, the magnetic fine particles, with the liquid medium being drawn hydrodynamically together with the particles. A schematic illustration of magnetic fluid is given in Figure 4.22. The magnetic powders are prepared by mechanical grinding or chemical synthesis and dispersed stably in an organic solvent by a dispersing agent like fatty acids.

As a liquid attracted by a magnet, magnetic fluid shows some interesting properties and has potential applications in many fields (Rosensweig, 1979; Charles and Popplewell, 1980). When the fluid is used as a vacuum seal of a rotating shaft, a magnet can prevent the liquid seal from spinning off and/or being sucked into the vacuum side. Sorting by means of specific gravity may be one of the most interesting applications of magnetic fluid: The magnetic fluid tends to come in the magnetic field and turn nonmagnetic materials out of the field. We can even separate the materials in terms of low and high density by a flotation method with increasing magnetic field. Many other potential applications of the fluid—such as for precise grinders, ink for ink-jet printers, speaker dampers, sound-wave absorbers, etc.—have been proposed, but few are currently used.

Antiscaling agent in desalination of seawater. Desalination of seawater is an important technology for providing drinking water in the countries of the Middle East. The desalination of seawater is done by either of two methods: distillation and reverse osmosis. In the distillation method, some kind of dispersant called an

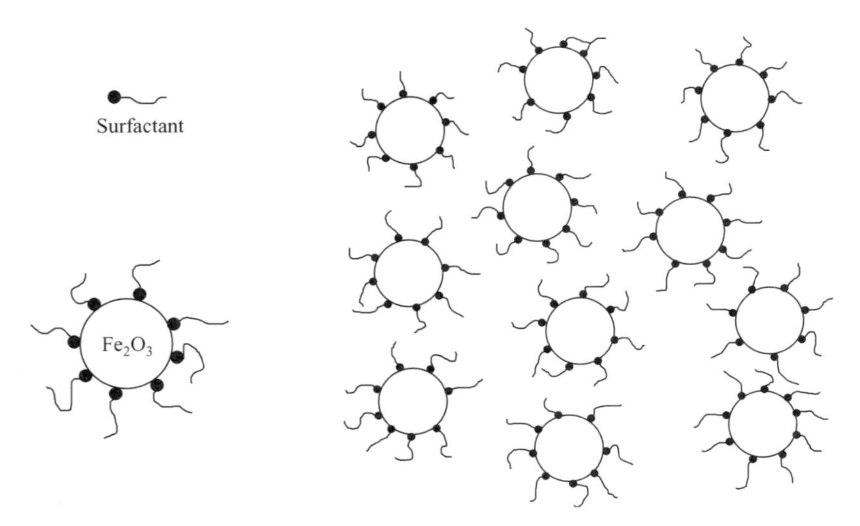

Figure 4.22 A schematic model of magnetic fluid. Magnetic powders are dispersed in an organic solvent and stabilized by surfactant adsorption.

antiscaling agent is essential. The antiscaling agent prevents scales of multivalent metal salts such as calcium and magnesium from depositing on the surface of a boiler. Sodium polyacrylate or a copolymer of sodium acrylate and sulfonate salt-type vinyl monomer is employed as the antiscaling agent.

Agricultural chemicals. Most agricultural chemicals—germicides, insecticides, herbicides, etc.—are water-insoluble organic compounds. These chemicals are sometimes used as a dispersion in an aqueous medium without dissolving them in an organic solvent; such dispersions are called *flowables*. The powders of such agricultural chemicals must be wetted with water and dispersed stably in the medium. Anionic dispersants such as alkylbenzene sulfonate, alkylnaphthalene sulfonate, and formaldehyde condensate of naphthalene sulfonate are used for this purpose. Nonionic dispersing agents such as polyoxyethylene alkyl ether and polyoxyethylene condensate of bis-phenol A are also employed.

Food. Dispersion is utilized also in some foods. For example, coffee creamer is a powdered milk that is dispersible in hot water; the dispersing agent is a kind of edible oil, monoglyceride. Some salad dressings contain some spices in a dispersed state, but a dispersing agent is not necessarily used, for example when the dressing is in a pasty liquid or is applied only after shaking.

b. Applications of Coagulating Action

Coagulation is also an important technology, and the basic principles on flocculation and coagulation were already mentioned in Section 3.2.4.*f*. When fine particles are in a dispersed state, an electrostatic repulsive force stabilizes the dispersion in many cases. The particles are mostly negatively charged in an aqueous medium. Therefore some cationic substances, which cancel out the negative charges present on the particles, are effective for coagulation. Cationic polymers and multivalent metal salts such as aluminum sulfate or alum are often employed. The cationic polymers or aluminum ions adsorb on the dispersed particles and diminish the negative charges on them. Multivalent metal ions also contribute to the coagulation by compressing the electric double layers through high ionic strength.

The bridging mechanism of coagulation (or flocculation) also works in very high molecular weight polymers. In this mechanism, one polymer molecule adsorbs on the surfaces of more than two particles and flocculates them. Figure 4.23 shows a schematic representation of the bridging mechanism of particle flocculation. A high molecular weight polyacrylamide is sometimes used for this

Figure 4.23 High molecular weight polymers aggregating the colloidal particles by the bridging mechanism.

flocculation purpose. Because a molecule of a very high molecular weight polymer has a length of as much as 10 μm in an extended state, such a molecule could possibly bridge two or more colloidal particles.

The most typical application of the coagulation technology is to precipitate the dispersed particles and eliminate them from a liquid medium. Water treatment in water-supply and drain systems is an example. The precipitation of active mud and reduction of its water content in wastewater treatment is one more example of the use of coagulating agents.

Fixers in the papermaking process are also a kind of coagulating agent. Several kinds of additives—such as fillers, dyes, and sizing agents—are used in papermaking. They must be fixed in the paper as much as possible, and so the fixers for them are important. To make paper, pulp fibers dispersed at about 1 wt % in water are layered on a fine net and then dried. Sometimes fillers such as talc and calcium carbonate are put inside the pulp layer to obtain high quality white papers. Because the filler particles are fine enough to pass through the net mesh, they must be preattached to the pulp fibers so they will not. The fixers are then hetero-coagulating agents of the fillers to the pulp fibers. Cationic starch and high molecular weight polyacrylamide are frequently used for these filler fixers.

Another important additive in papermaking is the sizing agent, which gives the paper appropriate water-repellency and is used to control the water absorption late in the papermaking process and in the final product. The most common sizing agent is rosin (main component: abietic acid). There are two types of rosin sizing agent: an aqueous solution of the sodium salt of the rosin and a colloidal dispersion of water-insoluble acid-type rosin. Both rosin sizing agents are deposited on the pulp fibers by the addition of aluminum sulfate. The water-soluble rosin of sodium salt becomes insoluble due to the formation of aluminum salt. The colloidal rosin is attached to the pulp by the coagulating action of the aluminum ions (as mentioned above). The aluminum sulfate solution is acidic (pH~4), so papers made in this way are not suitable for long-term preservation. A sizing agent that can be used in a neutral condition is provided as a colloidal suspension of alkylketene dimer (a kind of wax: see Section 3.2.1.*d*); the fixers of this neutral sizing agent are cationic polymers like cationic starches.

COM (coal–oil mixture) is a unique example of the application of coagulation. COM is a fuel in which fine powders of coal are mixed with heavy petroleum oil. The coal powders are often believed to be dispersed in the oil, but in reality they are coagulated lightly to form a network structure that prevents the coal particles from sedimenting to the bottom. High molecular weight polymers are utilized for the above coagulating agents.

c. Other Applications

Surfactant applications at solid/liquid interfaces are not limited to the interfaces between fine particles and liquids. The agents are also applied to the interfaces of bulk scale. The rust-preventive agent for metals is a typical example. Metals are corroded by rust formation, and so tremendous amounts of metallic resources can be saved if rust formation can be prevented. At present, the most effective way to protect metals against corrosion is to paint them, but this method cannot be applied in some cases. Plating with a rust-tolerant metal or a formation of a thin oxide layer (the passive state) is also a useful way. The easiest method for rust prevention, however, may be treatment with certain surfactants. Surfactant molecules adsorb on the metal surface, orienting their hydrophilic heads to the metal. Their hydrophobic tails then cover the metal surface and protect it against rust formation. Both oil-soluble and water-soluble rust-preventive agents are available. Fatty amines, the salt of fatty amine with fatty acids or alkylphosphates, sulfonated petroleum, etc. are the oil-soluble agents. Metal surfaces are coated with mineral oils containing these rust-preventive surfactants. This type of agent is employed for the storage of steel plates, in oil tankers, and so on. The water-soluble agents are alkanolamide of fatty acids, polyoxyethylene alkylamine, fatty amine acetate, alkylphosphates, etc. They are utilized in acid cleaning solutions for metals, water coolers, desalination plants, and so on.

Metal rolling oils are used to reduce the friction between a metal plate and a roll surface. Lubricating action under very high pressure is required for the rolling oils. Fatty acids, fatty acid esters, fatty alcohols, etc. are added to the rolling oils to achieve this. These surface active materials may adsorb on the metal surface with the same molecular orientation as that of rust-preventive agents. The hydrophobic chains covering the metal surface prevent direct contact between the metal and roll surfaces.

4.3.4 APPLICATIONS OF SOLID/GAS INTERFACE ADSORPTION

a. Lubrication

Hair conditioners and fabric softeners are applications of surfactant adsorption at solid/gas interfaces. They work as a lubricant for hair and fabrics, giving them a soft feeling. The hair conditioners lubricate the hairs in both wet and dry conditions. If the lubrication of hair is not good enough, the hairs are frictional and entangled, especially in long hair. When such entangled hairs are combed, they are damaged or are even torn off in the most serious case. Thus, hair conditioners are used not only for softness but also to protect hair from frictional damage.

Fabric softeners are used to make clothing feel soft. Cationic surfactants of the quarternary ammonium type are utilized for both hair conditioners and fabric softeners. Dialkyldimethyl ammonium salts and branched-chain monoalkyltrimethyl ammonium salts are preferable. It is often mentioned in textbooks that these cationic agents adsorb on hair or fabric surfaces orienting their cationic group toward the surfaces, since the substrates are negatively charged in water. Hydrophobic chains then cover the hair or fabric surface and prevent direct contact between hair or fabric fibers to give lubrication to them. The double-chain and branched-chain surfactants, however, form multilamellar vesicles in water (as described in Sections 3.3.3.*a* and 3.3.4.*b*). Thus, it is not easy to imagine how and why the molecules of the cationic agents in the multilamellar vesicles of onion-like structure adsorb on the hair or fabrics as a monolayer. Further studies are necessary on the lubrication mechanism of these cationic surfactants.

Dialkyldimethyl ammonium chloride is a very effective surfactant particularly for fabric softeners and has been used extensively for a long time. But since this surfactant is not easily biodegraded, new cationic surfactants for softeners are now being studied for development. Figure 4.24 shows some candidates of biodegradable cationic surfactants for fabric softeners. All contain ester bond(s) in their molecules and can be hydrolyzed in environmental conditions: Hydrolyzed compounds are biodegradable. In addition to biodegradability, the surfactants must show little or no toxicity for small living things in water. The water flea is commonly used to test toxicity. Unfortunately, there are no decisive com-

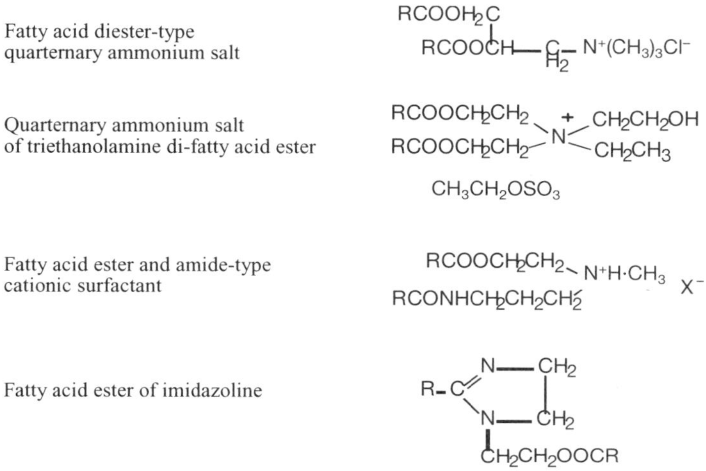

Fatty acid diester-type quarternary ammonium salt	$RCOOH_2C$ / $RCOOCH-CH_2-N^+(CH_3)_3Cl^-$
Quarternary ammonium salt of triethanolamine di-fatty acid ester	$(RCOOCH_2CH_2)_2N^+(CH_2CH_2OH)(CH_2CH_3)$ $CH_3CH_2OSO_3$
Fatty acid ester and amide-type cationic surfactant	$RCOOCH_2CH_2$, $RCONHCH_2CH_2CH_2$ $N^+H\cdot CH_3$ X^-
Fatty acid ester of imidazoline	$R-C$ (imidazoline ring: $N-CH_2$, $N-CH_2$) CH_2CH_2OOCR

Figure 4.24 Chemical structures of cationic surfactants expected to be candidates for new biodegradable fabric softeners.

pounds at present that satisfy all these required conditions, so research and development actively continues.

b. Antistatic Effect

Plastic resin and synthetic fabrics are easily charged, especially in wintertime. The charges on the plastic goods attract dusts, and the plastic surface is easily soiled. The charge on the synthetic fibers may cause us a mild electric shock when we touch a door knob after walking on a carpet or the unpleasant crackling sound of spark discharge when we take off a sweater. Antistatic agents are employed to depress such disadvantages as these. When antistatic agents are incorporated into the plastics during processing, the incorporated surfactant (antistatic agent) molecules migrate to the surface of the plastics and work at the solid/gas interfaces. A thin surfactant layer formed at the surface of the plastics sorbs water molecules from the atmosphere, and the sorbed water layer is believed to release the electric charges. Polyoxyethylene alkylamine, alkanolamide of fatty acids, alkyl sulfonates, etc. are used as this type of antistatic agent. Quarternary ammonium salts and betaine-type or imidazoline-type surfactants are very effective, but are not heat-tolerant. Thus, they are used only for external applications where the agent is coated on the surface of the plastics.

Antistatic agents for home use are also available. One type is an aerosol spray applied directly to clothing to keep it from adhering to the body. Cationic surfactants of the quarternary ammonium type are the main component of this product. Fabric softeners also have an antistatic property.

4.3.5 APPLICATIONS OF MULTIPLE INTERFACE ADSORPTION

The major application in this category is in the wetting control technologies. Antifogging agents for greenhouses and automobile window glass as well as the spreaders for agricultural chemicals are used to enhance the wetting of water or aqueous solutions. Water-repellent treatment is, of course, made to reduce the wetting of water. The action mechanisms of these treatments are shown in Figure 4.25.

a. Applications for Wetting Enhancement

Polyvinyl chloride and/or polyethylene films which possess low surface energy are used for agricultural greenhouses. Water is condensed on the films as droplets because of their low surface tension. Small water droplets attached on the films scatter and obstruct the sunlight, which should reach the ground and plants. To prevent this, antifogging agents are incorporated into greenhouse films during the manufacturing process. The agent molecules bleed out of the plastic films and decrease the interfacial tension of solid (film)/water. Part of the agents dissolve into water droplets and depress the droplets' surface tension. Reduction of both tensions contributes to the wetting of water on the film surface, as illustrated in Figure 4.25(a), and the thin water layer spread on the film keeps the sunlight from passing through. Polyoxyethylene sorbitan fatty acid esters (TWEEN), polyglycerin fatty acid esters, and mono- and/or di-glycerides are used for the antifogging agents of greenhouses.

Automobile glass surfaces should be able to be wetted with water because of the high surface tension of glass. But practically, it is not wetted well since most of the glass surface is usually soiled with oily dirt. When a surfactant solution is sprayed on the window glass of a car, the wetting of water is enhanced in the same way as described above.

Plant leaves sometimes have a number of prickles, which makes them very water-repellent (see Section 3.2.1.*d*). Thus, aqueous solutions or emulsions of agricultural chemicals cannot attach to the leaves when sprayed. Spreaders are then employed to wet the leaf surfaces with water so that the chemicals can attach to them. Mixed surfactant systems of nonionic agents such as polyoxyethylene alkyl ether and anionic agents such as alkylbenzene sulfonate are frequently utilized as spreaders for agricultural chemicals.

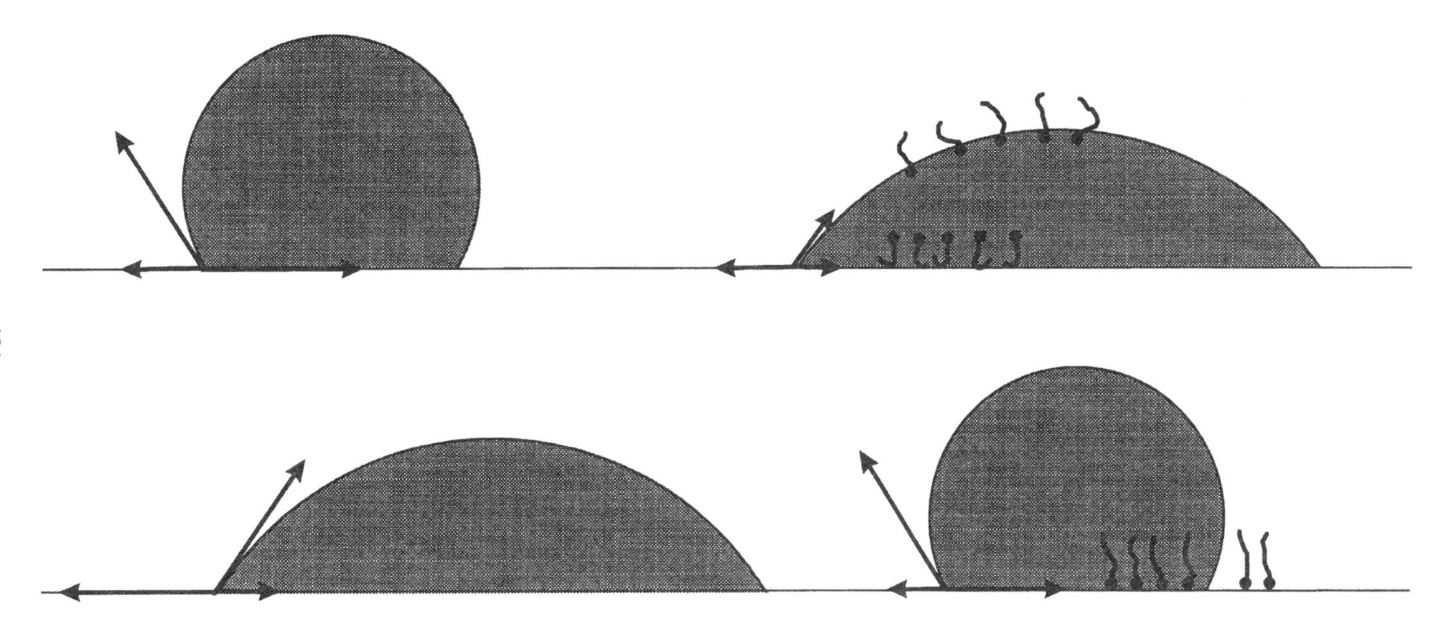

Figure 4.25 Action mechanism of wetting control agents. Shown here are a wetting enhancer (like antifogging agents and spreaders for agricultural chemicals) (a) and a water- and/or oil-repellent treatment (b).

b. Water- or Oil-Repellent Treatments

Inorganic powders for cosmetic use are commonly treated with certain surfactants to give a water- and/or an oil-repellent surface. Cosmetic powders for, say, foundations must be strong against sweat and oily fat, and require this treatment. The principle of the water- and/or oil-repellent treatment is to alter the high surface energy of the substrate to a lower one by covering the surface with some surfactant molecules. As shown in Figure 4.25(b), the surface tension of the solid substrate becomes low and the interfacial tension between liquid and solid increases by this treatment. The contact angle then also increases. The lower the surface tension of the treated surface, the more effective the water- and/or oil-repellency. So fluorinated or silicone-type surfactants are frequently employed.

In addition, the surfactant molecules should adsorb orientating their hydrophilic head toward the solid surface, as illustrated in Figure 4.25(b). Phosphate groups that interact strongly with the metal ions of a solid surface are chosen as a hydrophilic head group to obtain this molecular orientation. Silane or titanate coupling agents are also powerful for water-repellent treatments. They react with the —OH group on the solid surface to form covalent bonds. Cationic surfactants are useful for a negatively charged surface since their molecules adsorb orienting the cationic head to the solid surfaces. The water-repellent treating agents are also applied to umbrellas, raincoats, and so on.

c. Other Applications

Adjuvant for agricultural chemicals is used to enhance the efficacy of the chemicals. When used together with the adjuvant, herbicides and insecticides function in smaller amounts or more effectively in normal amounts. The action mechanism of the adjuvant is not clear yet. It may be multiple effects at work: a wetting action like a spreader, acceleration of the chemical absorption into plant tissues, and the improvement of solubility and/or dispersion of the chemicals. In any case, the adjuvant must work at multiple interfaces.

4.4 Applications Utilizing Aggregation Phenomena

Aggregation phenomena such as micelle and/or vesicle formation are one of the most characteristic properties of surfactants. Unexpectedly, however, the direct applications of the aggregation phenomena of surfactants are quite few compared with those of the adsorption phenomena. In other words, the possibility of the applications of aggregation phenomena is expected in the future, and some of them are mentioned in this section.

4.4.1 APPLICATIONS OF MICELLAR SYSTEMS

a. Emulsion Polymerization

Emulsion polymerization is practically the only major application of surfactant micelles in industry. As pointed out in Section 4.3.2.*d*, the micelles are essential in emulsion polymerization technology even though the process is termed *emulsion polymerization.* Polymerization reactions proceed in micelles in emulsion polymerization, and the emulsion droplets of monomer are just the reservoir of the monomer supply. The classic theory for emulsion polymerization is the Smith-Ewart theory (Smith and Ewart, 1948), and we will start this section with their theory.

Figure 4.26 shows a schematic illustration of the process of the emulsion polymerization proposed by Smith and Ewart. This process can be typically applied to hydrophobic monomers like a styrene. Monomer molecules are solubilized into surfactant micelles (stage (A) in the figure). Radicals of an initiator are generated in an aqueous phase and come into some micelles in which the monomer molecules are present (B). The polymerization reaction starts in the micelles by initiation of a radical (C) and terminates when the other radical comes in again (D). The polymerization process is completed, and a polymer molecule is synthesized (E). The monomer molecules, supplied from emulsion droplets of the monomer, are solubilized again into the micelle containing a polymer molecule.

The polymerization starts again when another radical comes into the micelle from the aqueous phase, and the same process repeats. This process of emulsion polymerization was quantitatively formulated by Smith and Ewart on the assumptions that (1) all the radicals generated in the aqueous phase are trapped in the surfactant micelles, and (2) the polymerization and termination reactions take place only in the micelles and not in the aqueous phase or in the monomer droplets (Smith and Ewart, 1948). These assumptions are reasonable because of much larger specific surface area of the micelle than that of the monomer droplet and are validated almost exactly when the monomer is hydrophobic.

Suppose that there are N reaction places (micelles containing monomer molecules) in 1 ml of the solution and that n radicals are present in one of the micelles

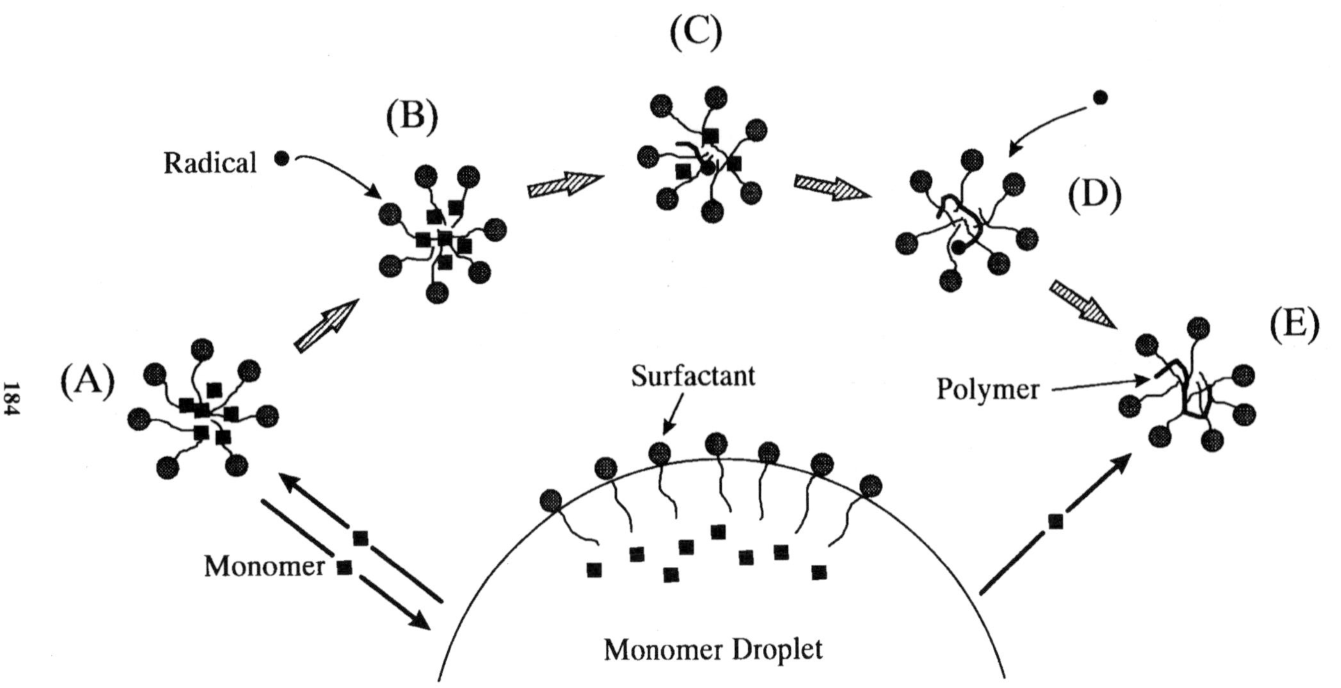

Figure 4.26 Schematic representation of the emulsion polymerization process. Stage (A) is the solubilization of monomer molecules into micelles, (B) is the radical entrapment in the micelles, (C) is the start and progress of the polymerization reaction, (D) is the termination of the polymerization reaction by another radical, and (E) is the completion of the reaction.

examined. Then the rate of the radical number change in the micelle to be watched can be written as

$$\frac{dn}{dt} = \frac{\rho}{N}, \tag{4.7}$$

where ρ is the generation rate of radicals in 1 ml of the solution. Eq. (4.7) implies that all the radicals generated in the aqueous phase are trapped in the micelles, as assumed above. The escape rate of the radical once incorporated into a micelle is proportional to the concentration of the radical in the micelle and the surface area of the micelle. Thus,

$$\frac{dn}{dt} = -k_0 s \frac{n}{v}, \tag{4.8}$$

where s and v are the surface area and the volume of the micelle, respectively, and k_0 is the rate constant of radical escape. The rate of radical disappearance in a micelle by the combination of two radicals can be expressed as

$$\frac{dn}{dt} = -k_t \frac{n(n-1)}{v}, \tag{4.9}$$

where k_t is the termination rate constant of the radicals. Eq. (4.9) is easily understood by noting that the combination number of selecting two radicals from n radicals in the micelle is $n(n-1)/2$, and two radicals disappear by one chance of the combination.

Let us move on to consider the entire system from the one reaction place (micelle). Suppose that N_n micelles containing n radicals are present in 1 ml of the solution, and the total number of micelles is expressed as

$$N = N_0 + N_1 + N_2 + \cdots + N_n + \cdots. \tag{4.10}$$

If we consider a steady state attained after the polymerization reaction proceeds to some extent, we obtain

$$N_{n-1}\frac{\rho}{N} + N_{n+1}k_0\frac{s(n+1)}{v} + N_{n+2}k_t\frac{(n+2)(n+1)}{v}$$
$$= N_n\frac{\rho}{N} + N_n k_0\frac{sn}{v} + N_n k_t\frac{n(n-1)}{v}. \tag{4.11}$$

This equation means that the number of micelles containing n radicals (N_n) is kept constant by the balance of the rates of generation and disappearance of the micelle.

If we could solve Eq. (4.11) exactly, N_n's would be obtained as a function of n and the reaction rate of polymerization could be expressed in general. Since it is not possible to solve it exactly, the equation is solved here approximately; approximate solutions give us enough information to gain insight into the essential points of emulsion polymerization. The following conditions are reasonable for solving the equation approximately, taking the practical application of the emulsion polymerization into consideration.

1. A radical once trapped in a micelle never does escape from the micelle ($k_0 = 0$).
2. The combining rate of two radicals in a micelle is much faster than the entering rate of a radical from the aqueous phase ($\rho/N << k_t/v$).

Eq. (4.11) can be rewritten as Eq. (4.12) from the first condition:

$$N_{n-1}\frac{\rho}{N} + N_{n+2}k_t\frac{(n+2)(n+1)}{v} = N_n\frac{\rho}{N} + N_n k_t\frac{n(n-1)}{v}. \tag{4.12}$$

Furthermore, from the second condition, the number of micelles containing more than two radicals must be very few. Then, we have only the terms $n = 0$ and $n = 1$ of Eq. (4.12).

$$N_2 k_t\frac{2}{v} = N_0\frac{\rho}{N} \qquad \text{for } n = 0 \tag{4.13}$$

$$N_0\frac{\rho}{N} + N_3 k_t\frac{6}{v} = N_1\frac{\rho}{N} \qquad \text{for } n = 1 \tag{4.14}$$

We may be able to put $N_3 \cong 0$ from the second condition. Then we obtain

$$N_0 \approx N_1. \tag{4.15}$$

We can rewrite Eq. (4.13) as

$$N_2 = \frac{\rho/N}{2k_t/v}N_0.$$

From the second condition, we obtain

$$N_2 << N_0. \tag{4.16}$$

The total number of radicals can be expressed as

$$n_T = \Sigma n N_n = N_1 + 2N_2 + 3N_3 + \cdots \tag{4.17}$$

Finally we obtain an important result of Eq. (4.18) by combining Eqs. (4.10), (4.15), and (4.16) and putting $N_3 \cong 0$:

$$n_T = \frac{N}{2}. \tag{4.18}$$

This result indicates that half of the micelles (reaction places) contain one radical and the remaining half are empty. This conclusion is one of the most important results of the Smith-Ewart theory.

Some of the results characteristic of emulsion polymerization are derived from Eq. (4.18). For example, the rate of polymerization reaction in 1 ml solution, R_P, is expressed as

$$R_P = k_P \frac{N}{2} [M], \tag{4.19}$$

where $[M]$ is the monomer concentration in the reaction place (micelle) and k_P is the rate constant. The reaction rate depends chiefly on the number of reaction places, N, since the monomer concentration $[M]$ is virtually kept constant by the monomer supply from the reservoir (emulsion droplets) during the steady-state polymerization reaction. In addition, the polymerization rate does not depend on the rate of radical generation in the aqueous phase, because no reaction takes place in water. The average degree of polymerization, p, is determined by the interval time between the arrival of an initiating and of a terminating radical at a reaction place (micelle) and is expressed as

$$p = k_P \frac{N}{\rho} [M]. \tag{4.20}$$

The interval time is the reciprocal of hitting frequency of the radical at a micelle and is equal to N/ρ. The chain extension of a polymer within the interval time is expressed by Eq. (4.20), since the chain extension rate of a polymer is $k_P[M]$.

From the industrial point of view, emulsion polymerization is extensively used in manufacturing the synthetic rubbers, resins, etc.; it is a simple process requiring inexpensive production facilities. However, because a surfactant remains as a contaminant in the products of the emulsion polymerization, this method is not suitable in some cases, especially for products requiring water-tolerance. One advantage of this technique is that beautiful monodispersed spherical particles can be produced. Figure 4.27 is a scanning electron microscope (SEM) photograph of polystyrene latex. The particle size of the latex is in the range of submicrometer, and the regular structure—like a crystal—of the particles shows beautiful iridescence due to the diffraction of visible light. Another advantage is

Figure 4.27 SEM image of monodispersed polystyrene latex particles synthesized by the emulsion polymerization technique. The particle diameter is submicrometer in size, and the ordered structure of the particles shows beautiful iridescence, like an opal. The length of the bar in the figure is 1 μm.

that the heat of polymerization can be easily dissipated (compared with bulk polymerization). This advantage is also present in suspension polymerization. The main surfactant used in emulsion polymerization is still soaps, although anionic surfactants such as linear alkylbenzene sulfonates, alkyl sulfates, formaldehyde condensates of naphthalene sulfonates, di-sulfonates of alkyl diphenyl ether, etc. and nonionic surfactants such as polyoxyethylene alkyl ether are also employed.

b. Solubilization and Detergency

Solubilization of oils into surfactant micelles is sometimes proposed to contribute to the detergency in dishwashing and dry cleaning processes. But this author thinks the contribution of solubilization may be too small. One spherical micelle can normally solubilize several oil molecules, i.e., about 100 surfactant molecules take several oil molecules into the water phase. Thus, the rolling-up and emulsifying mechanisms make a much greater contribution to the cleansing of oily dirt. From the same reasoning, secondary solubilization of water-soluble soils into solubilized water may also show little effect on the detergency in dry cleaning. In microemulsion regions, however, a large amount of oils or water can be incorporated into the systems. Although not yet utilized, microemulsion is probably a good candidate for future liquid detergents.

4.4.2 APPLICATIONS OF LYOTROPIC LIQUID CRYSTALS

a. Liquid Crystal Emulsification

In usual emulsions, a monomolecular surfactant layer adsorbed on the liquid droplets prevents the drops from coalescence and stabilizes the emulsion. In the liquid crystal emulsification, on the other hand, it is the higher-order structure of the liquid crystal that stabilizes the emulsion. Thus, the emulsions formed by the liquid crystal are quite stable and very useful in practical applications.

Liquid crystal emulsification provides some very unique emulsions. They can be W/O (O/W) emulsions containing 80–90 wt % of water (oil) in which the liquid of much smaller amount forms the continuous phase. Let us consider first the W/O emulsion containing 80–90 wt % water. The stability of an emulsion made with α-mono-isostearylglyceryl ether (GE) was compared with those stabilized by usual surfactants, as shown in Figure 4.28 (Suzuki and Tsutsumi, 1987). In the GE emulsion, no water separation is observed and the oil separation becomes smaller with the increasing amount of oil. In such W/O emulsions as this, the water droplets occupying more than 90 wt % must be packed very closely and the thin oil layers that separate them must be tough enough to prevent the closely packed water droplets from coalescing. Liquid crystal contributes to form the tough oil phase.

The GE of liquid state absorbs water and forms a liquid crystal of reversed middle phase in the range of 8–23 wt % water content. The water further added is separated out of the liquid crystal. When an oil phase is added to this GE/water system, the phase diagram shown in Figure 4.29 is obtained. The emulsion demonstrated in Figure 4.28 is represented by the dotted line in Figure 4.29, since the GE concentration is kept constant at 5 wt %. The stable region of the emulsion is located at higher water content on this dotted line and is inside the curve CAW. One can understand that the stable composition of the emulsion is the two-phase region of liquid crystal and water. The oil added is completely in-

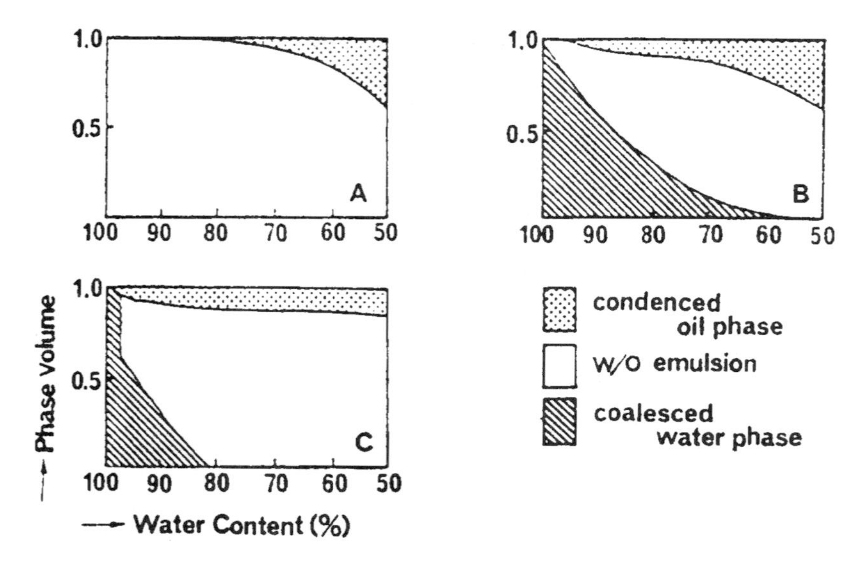

Figure 4.28 Comparison of the emulsion stability prepared with GE and common emulsifiers (Suzuki and Tsutsumi, 1987). Emulsifier concentration: 5 wt %. Data obtained at 25°C for 7 days.

corporated in the liquid crystalline phase and not present as a separated liquid oil phase. Thus, we have to call this emulsion a W/LC (water in liquid crystal) type rather than a W/O one. The oil incorporated in the liquid crystal is never separated out, and the continuous phase made of liquid crystal is tough enough to prevent the coalescence of water droplets. This is why the liquid crystal emulsification provides such stable emulsions. When applied in skin care creams or creamy foundations, the above emulsion of high water content gives no oily feeling in spite of the W/O emulsion.

Now let us consider the O/W emulsion containing more than 80 wt % oil phase (Suzuki *et al.*, 1989). Stating the conclusion first, this emulsion is also an O/LC (oil in liquid crystal) type, not a simple O/W emulsion. The surfactant that gives the O/LC emulsion is mono L-arginine salt of β-branched monoalkyl phosphate (hexyldecyl phosphate). Polyol (glycerol preferably) is also essential although its role is not clear. Another characteristic property of this emulsion is that any kind of oil can be stabilized by this liquid crystal regardless of its re-

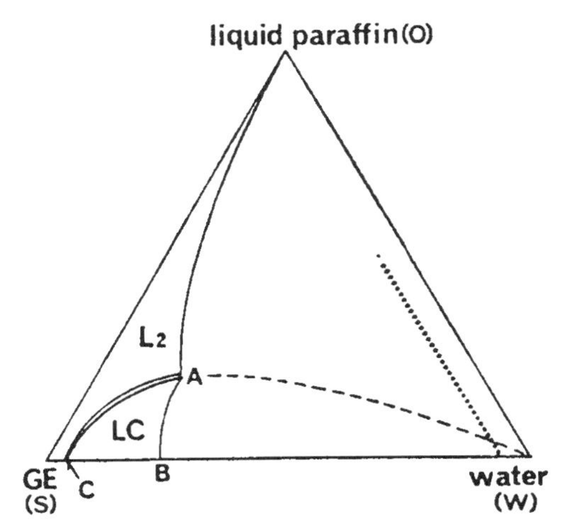

Figure 4.29 Ternary phase diagram of GE/water/oil (liquid paraffin) system at 25°C (Suzuki and Tsutsumi, 1987). LC: liquid crystalline phase; L_2: isotropic solution.

quired HLB value. The very existence of the O/LC emulsion itself substantiates that the liquid crystal does not dissolve into the oil. In addition, the surfactant (L-arginine hexyldecyl phosphate) is not soluble in water and forms a lamellar liquid crystal in a wide range of its concentration. Consequently, the oil droplets can be dispersed stably, being surrounded by the liquid crystal even when the O/LC emulsion is diluted with water (Suzuki *et al.*, 1989). Interestingly, the droplet size in the emulsion thus obtained depends on the oil/surfactant ratio.

Our final example of liquid crystal emulsification is a combination of common surfactants. Liquid crystal is formed by mixing a polyoxyethylene-type nonionic surfactant like TWEEN with a fatty alcohol in water. When the temperature is high, the fatty alcohol dissolves in the oil phase. But the solubility of the alcohol decreases with decreasing temperature, and the alcohol molecules bleed out and meet with the nonionic surfactant present in the aqueous phase to form a liquid crystal at the surface of the oil droplets. A schematic illustration of this emulsion is shown in Figure 4.30 (Suzuki *et al.*, 1984).

b. Liquid Crystal Cleanser

A lamellar liquid crystal is cleverly utilized as a cosmetics remover (Suzuki *et al.*, 1992). Most cosmetics consist of oils and other ingredients such as pigments

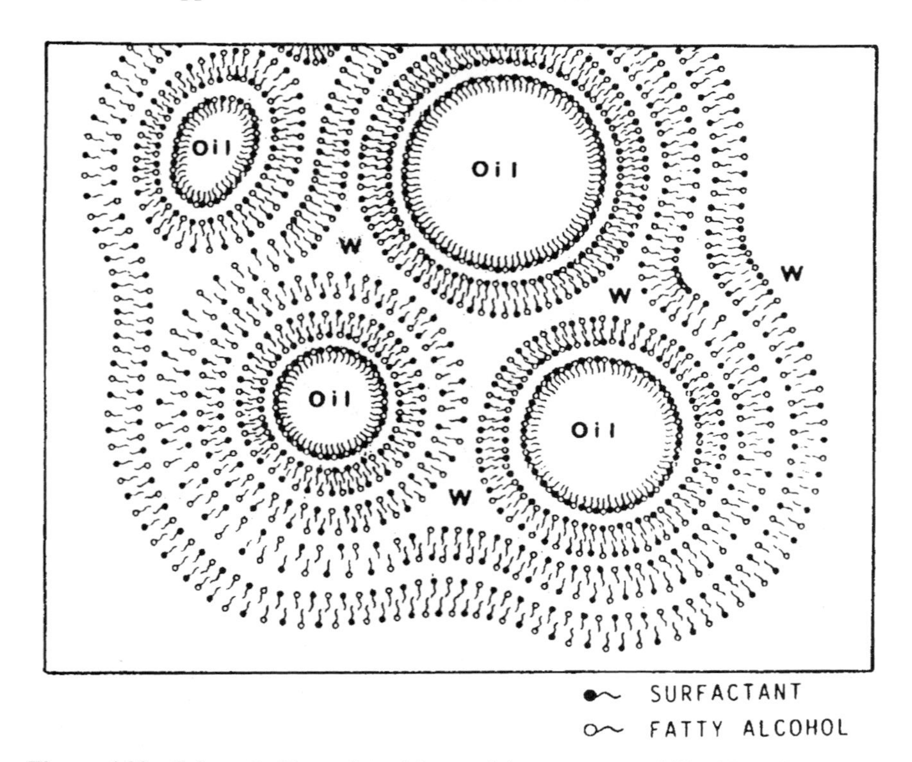

Figure 4.30 Schematic illustration of the emulsion structure stabilized by a liquid crystal consisting of surfactant and fatty alcohol (Suzuki *et al.*, 1984). Reprinted from *J. Dispersion Sci. Technol.,* **5**, 119 (1984) by courtesy of Marcel Dekker, Inc.

and inorganic powders. Thus, an oily component is essential in the cleanser to remove the cosmetics from the skin. But oily cleansers, like cleansing creams, are not easily washed off with water; they are usually wiped off with a tissue paper, which over time may irritate and damage the skin. An oily cleanser that can be easily washed off with water is thus more desirable, and a lamellar liquid crystal containing a large amount of oil serves this purpose.

Figure 4.31 shows a phase diagram of a pseudo-three-component system. The surfactant used is a branched nonionic one—polyoxyethylene (20 oxyethylene units as an average) octyldodecyl ether—and it easily forms a lamellar liquid crystal in water. A surfactant having a branched hydrophobic group behaves like a double-chain surfactant and readily forms a planar micelle by the packing

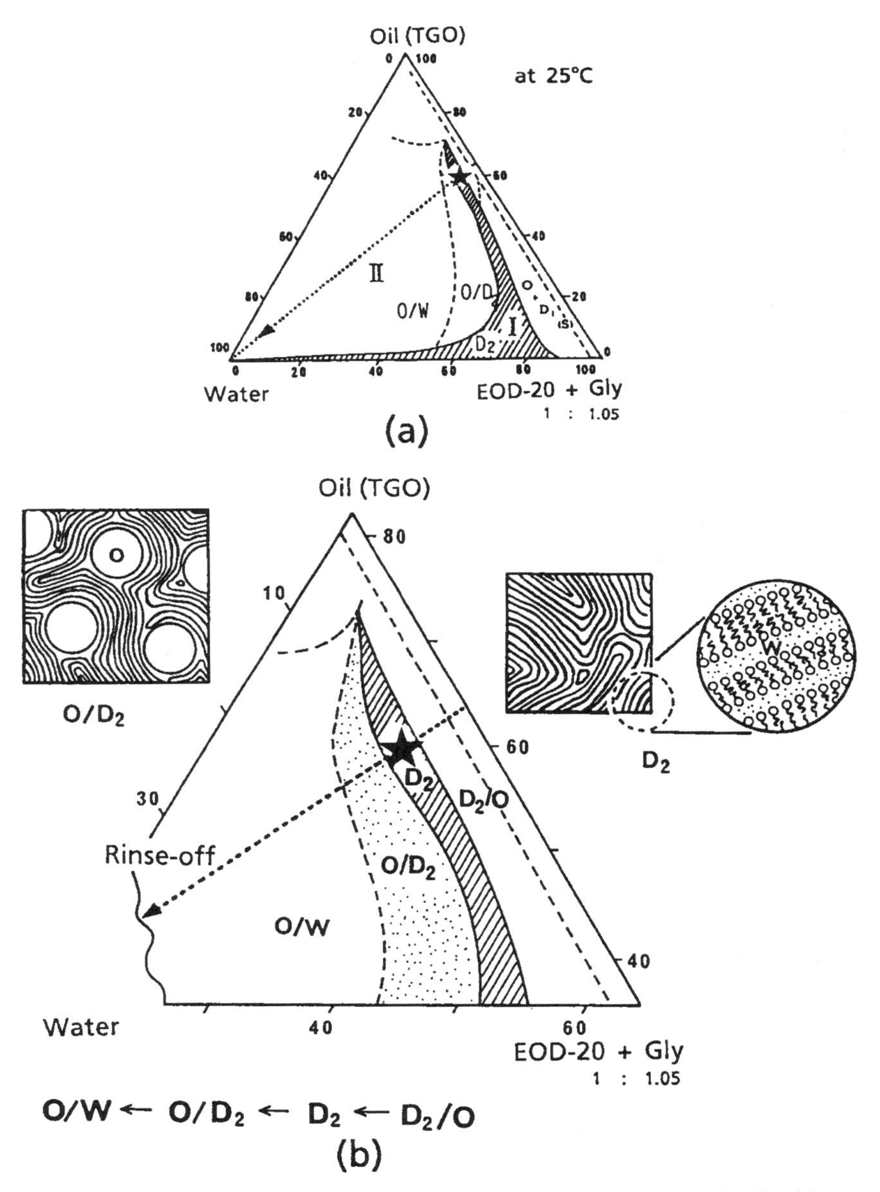

Figure 4.31 Phase diagram of the pseudo-three-component system of a liquid crystalline cosmetics cleanser (Suzuki *et al.*, 1992). The composition marked with a star (★) symbol is the cleanser formulation. Reprinted by permission of the Society of Cosmetic Chemists.

restriction (see Section 3.3.2.*b*). A large amount of oil can be incorporated into the lamellar liquid crystal, and a sharp peninsula of the liquid crystalline phase toward the oil corner can be observed in the figure. The cosmetics cleanser is formulated at the composition of a tip of the peninsula marked by a star symbol. The liquid crystal shifts to the two-phase region by excess oil when in contact with oily dirt (the cosmetics) and dissolves the dirt into released excess oils. The system comes back again to the liquid crystalline phase when in contact with water on rinsing, and finally it goes into the O/W emulsion region. Thus, the cosmetics can be easily washed off with water without scrubbing.

c. Mesoporous Materials Made with Liquid Crystal Templates

Lyotropic liquid crystals of surfactants possess unique and interesting structures in terms of materials development. The periodic hexagonal array of cylindrical micelles (middle phase), the regularly stacked bilayer membranes (lamellar phase), the infinite periodic minimal surfaces of bilayer membranes (bicontinuous cubic phase) and so on all seem to be quite useful if we can fix these structures in solid materials. This original idea has been realized by the liquid crystal template method (Kresge *et al.*, 1992; Beck *et al.*, 1992). Water-soluble raw materials of silicate (or aluminosilicate) are filled in the aqueous phase of a surfactant liquid crystal and reacted in an autoclave. The as-synthesized inorganic material is calcined at high temperature. Organic surfactant is burned out on calcining, and the calcined silicate has mesopores (in which surfactant aggregates used to exist forming a liquid crystal). The hexagonal phase of cationic surfactant, hexadecyl trimethyl ammonium salt, was used as a template. Synthesis of essentially the same material has been reported by another research group, but they propose a different mechanism (folded sheet formation mechanism) than templating (Inagaki *et al.*, 1993, 1996). Figure 4.32 shows transmission electron microscope photographs of the silicate material synthesized by this folded sheet method. The beautiful ordered structure of the hexagonal and layered array can be seen from the photograph.

A generalized method to synthesize materials other than the hexagonal array has been also proposed (Huo *et al.*, 1994). The proposed principle of templating is very simple, but the actual mechanism is still in controversy.

The pore size of a mesopore is between 2 nm and 50 nm; a pore smaller than 2 nm is a micropore and one larger than 50 nm is a macropore. Mesoporous material is expected to be useful in the fields of catalysts, adsorbents, molecular sieves, and so on, but none practically applied in any industries yet.

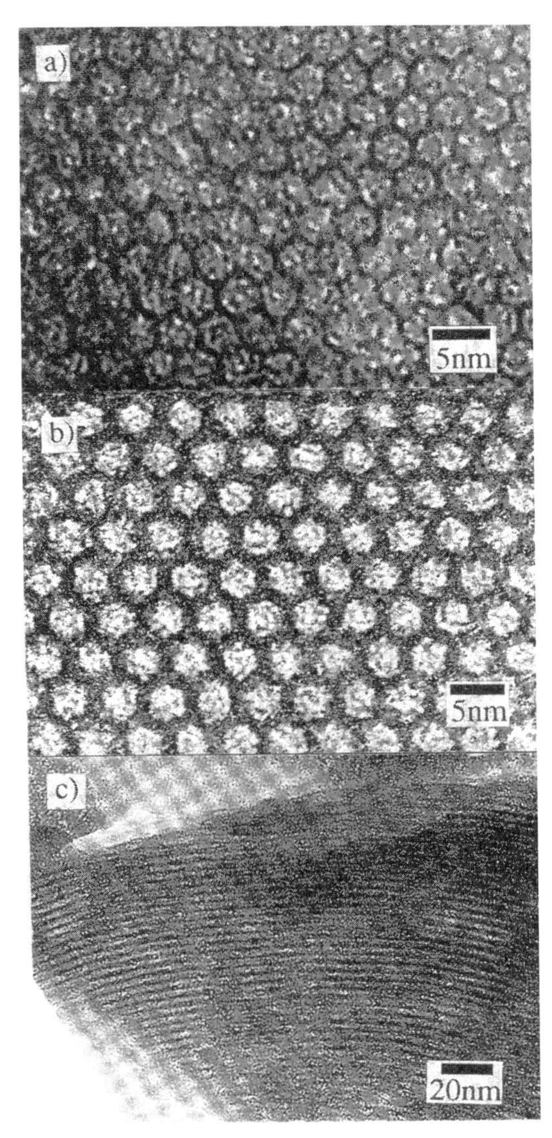

Figure 4.32 Transmission electron microscope photographs of a mesoporous silica material prepared by the folded sheet formation mechanism, utilizing an ordered structure of surfactant liquid crystal (Inagaki *et al.*, 1996).

4.4.3 APPLICATIONS OF BILAYER MEMBRANES

The bilayer membrane, including vesicles and liposomes, is a quite unique material and works somewhat like a biomembrane. Thus, it possesses superior potential for a wide variety of applications, especially in the medical field. But few applications have been made so far, and extensive applied research is required.

a. Drug Delivery Systems (DDS)

An effective drug delivery system has been a major goal of liposome technology from its beginnings. A drug carrier targeting an affected part of the viscera is the typical DDS. The ultimate goal is that the drug be carried only to the affected part and not to any healthy viscus, i.e., that there are no side effects. In the early stages of the research, trials to obtain the targeting for viscera were done by controlling the size, surface charge, chemical structure of lipid, cholesterol content, etc. of liposomes. All such trials, however, have not been successful so far. Thus, an active targeting method utilizing molecular recognition has been developed instead. The active targeting ability of liposomes means that the drug-bearing liposomes can search out the affected part of a human body by themselves and release the drugs there. These liposomes must recognize the affected cells and possess the molecular recognition ability. Two mechanisms of molecular recognition have been discovered so far. One is the interaction between saccharide chains, and the other is the antigen-antibody interactions. An example of saccharide chain interaction was already mentioned in Section 3.3.4.*c*, so we will consider an example of antigen-antibody interactions here.

Targeting drugs to cancer cells is one of the most important problems to be solved in the present medical field, and the antigen-antibody interaction may be the best way for DDSs to recognize the cancer cells. A monoclonal antibody is essential to exactly distinguish the cancer cells from normal ones. A method of attaching the antibody to a liposome surface is important to obtain a stable immuno-liposome. An antibody directly bonded with a hydrophobic compound cannot be incorporated stably into the liposomal membranes probably because of the steric hindrance of the bulky antibody molecule. Some spacer may be necessary between the antibody and the hydrophobic moiety to obtain a stable immuno-liposome. Sunamoto *et al.* have succeeded in making such immuno-liposomes by employing the antibody hydrophobized with cholesterol using a pullulan chain as a spacer (Sunamoto *et al.*, 1987). The immuno-liposome thus obtained is successfully concentrated in the cancer cells.

b. Artificial Vaccine

A vaccine is a preparation of dead or deactivated germs that induces antibody production in a human body by injecting the vaccine or the antibody itself obtained by vaccination. Is the whole germ body necessary to obtain the vaccine activity? No. It has been found that liposomes bearing membrane proteins of a cancer cell are able to induce the antibodies for the cancer, thus working as a vaccine (Sunamoto *et al.*, 1990b; Shibata *et al.*, 1991; Sunamoto *et al.*, 1992b). The membrane proteins of a tumor cell, BALB RVD, were directly transferred to liposomes containing an artificial boundary lipid (1,2-dimyristoylamido-1,2-deoxyphosphatidyl choline: DDPC, see Section 3.3.4.*c*), and the liposomes so obtained were used as an artificial vaccine. This liposomal vaccine was injected twice into mice, and then a tumor cell was injected under the skin. To test the efficacy of the liposomal vaccine, the size of the growing tumor was measured. Figure 4.33 clearly shows the effectiveness of the artificial vaccine in reducing the tumor size (Sunamoto *et al.*, 1990b). It is not yet clear which proteins transferred from the tumor cells can immunize the mice. But this method may eventually be able to identify the protein(s) effective for immunity and may contribute to the medical research of cancers.

c. Stimuli-Responsive Membranes—Sensors.

Response to stimuli is one of the most important functions of biomembranes. For example, photostimulus to sight cells and chemical stimulus to taste and/or olfactory cells induce the impulsive response of nerve cells. Artificial membranes possessing a stimuli-responsive function similar to that of biomembranes have been developed. Such artificial membranes are expected to be applied to certain sensors and a stimuli-responsive drug-releasing device (see Section 3.3.4.*c*). Here, we discuss some model membranes for chemoreception, which mimick the taste and/or olfactory cell membranes, and their applications to sensors.

A microporous filter (Milipore filter) containing in its pores the lipids extracted from a cow tongue was employed as a model membrane for chemoreception (Kamo *et al.*, 1974a). Figure 4.34(a) illustrates an apparatus to measure the membrane potential appearing when taste substances are applied. Two parts of the cell are separated with the model membrane. One side of the cell is filled with a reference solution (300 mmol/l NaCl), and a taste substance is added in the other side of the cell. The membrane potential changes that occur with the addition of several taste stimuli are shown in Figure 4.34(b). These responses are quite similar to physiological ones of some animals. The membrane potential

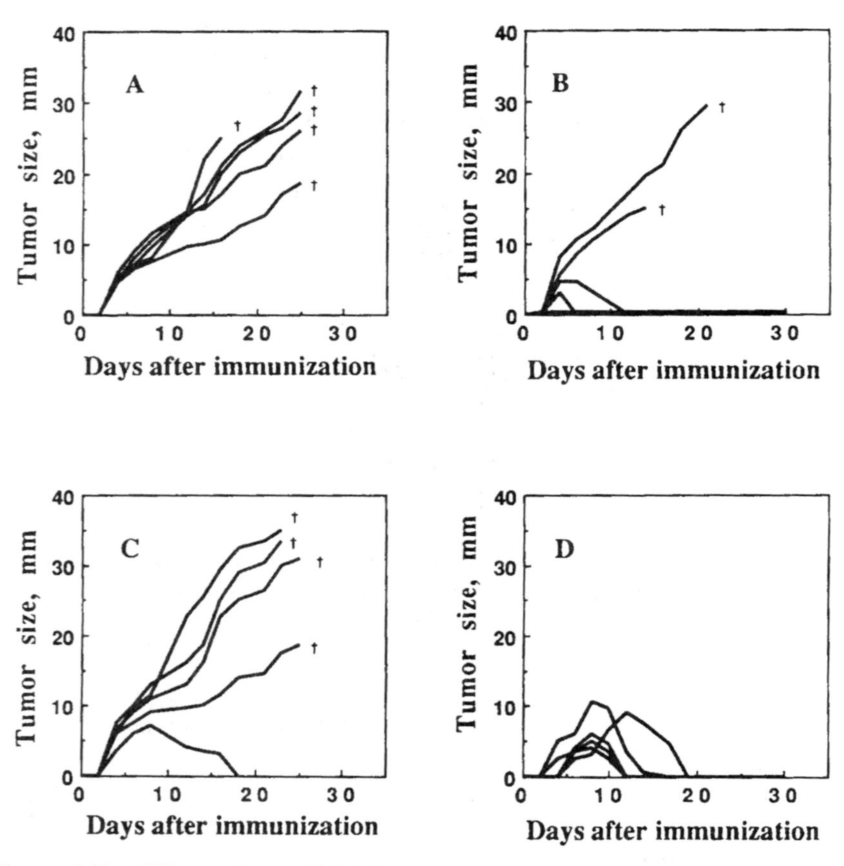

Figure 4.33 Efficacy of an artificial liposomal vaccine (D) in growth inhibition of a tumor (BALB RVD cell) (Sunamoto *et al.*, 1990b). (A)–(C) are controls.

change originates from the interfacial potential difference between two sides of the model membrane in contact with the solutions of reference and taste substance, and not from a diffusion potential due to ion permeation through the membrane (Kamo *et al.*, 1974b).

A liposome made of soybean lecithin was found to be responsive to odorous chemicals (Nomura and Kurihara, 1987a; 1987b). The membrane potential difference between the inner and outer aqueous phases of the liposome was measured using a special fluorescent dye. The membrane potential changes that occur with the addition of odorous chemicals depend on the concentration of the chemicals. Threshold concentrations of the odorants for the membrane potential

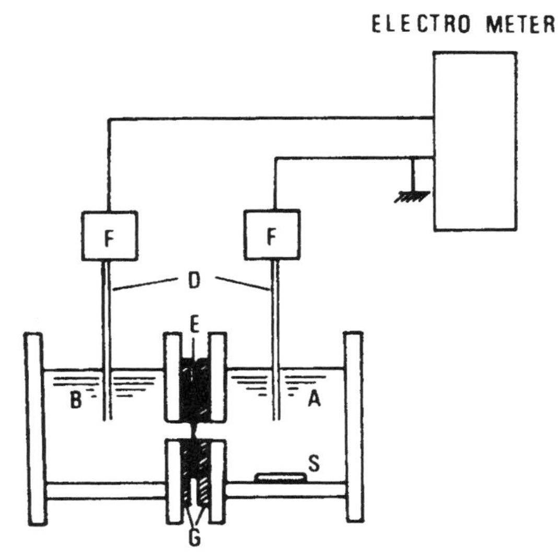

(a) A: compartment for stimulating solution, B: compartment for reference salt solution, D: saturated KCl salt bridges, E: model membrane, F: calomel electrodes, G: spacer of silicone rubber gasket, S: magnetic stirrer tip

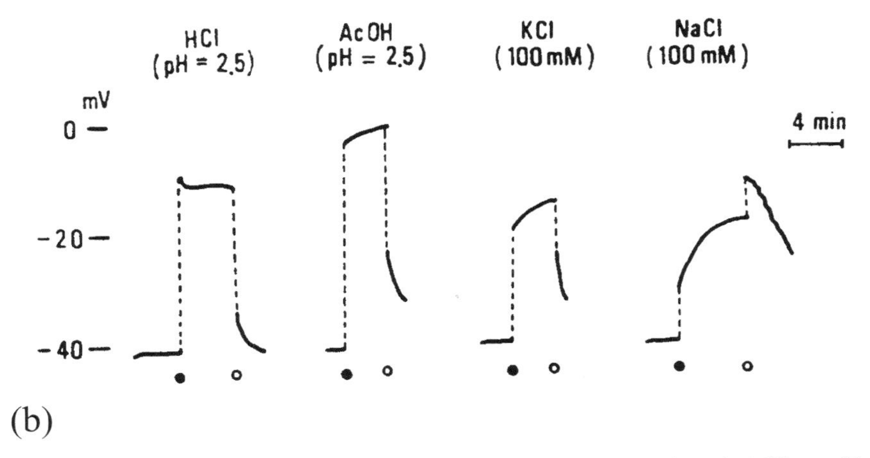

(b)

Figure 4.34 Response of a model membrane constructed with lipids and a Milipore filter to taste stimuli (Kamo *et al.*, 1974a). (a) Schematic representation of an apparatus to measure the membrane potential. (b) Membrane potential changes responsive to taste substances; the taste stimulus was applied at the point (•) and washed away at the point (○). Reprinted from *Biochim. Biophys. Acta,* **367,** 11 (1974) with kind permission of Elsevier Science—NL, The Netherlands.

change are compared with those of the olfactory cell of a frog and a pig in Figure 4.35. Both threshold values are in good relationship. It is very interesting and even surprising that the soybean lecithin liposomes having no relation to olfaction show the response similar to the olfactory cells.

Our final examples are membranes responsive to odorants or bitter taste substances made of a totally synthetic lipid (surfactant). A bilayer membrane of dioctadecyldimethyl ammonium (DOA) salt was coated on a piezoelectric quartz cell or used as a separating membrane, and was tested as a chemoreceptive model membrane (Okahata *et al.*, 1987; Okahata and En-na, 1987; Okahata and Shimizu, 1987). To prepare these thin DOA films, the polyion complexes of DOA with anionic polymers like a polystyrene sulfonate are utilized (Kunitake *et al.*, 1984; Higashi *et al.*, 1987). The polyion complexes dissolved in an organic

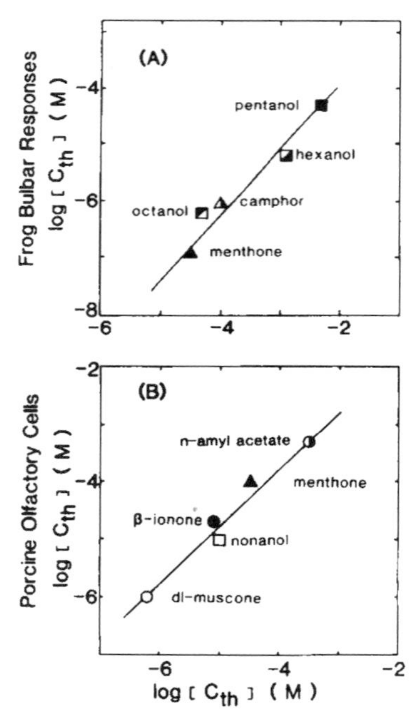

Figure 4.35 Comparison of the threshold values of odorant concentrations in the membrane potential between a liposome of soybean lecithin and a frog olfactory cell (a), and between the liposome and a pig olfactory cell (b) (Nomura and Kurihara, 1987a. Reprinted with permission from *Biochemistry,* **26,** 6135 (1987), American Chemical Society.

solvent (such as chloroform) are cast and dried. A thin film consisting of multiple bilayer membranes is obtained and can be used as a coating film on a piezoelectric quartz crystal or a self-standing separating membrane for potential measurements.

Figure 4.36(a) illustrates a schematic drawing of the apparatus of the DOA-coated piezoelectric cell. Some results obtained by this apparatus are shown in Figure 4.36(b). As is well known, the proper frequency of a piezoelectric quartz cell decreases linearly with the increment of mass attached to the cell surface. A

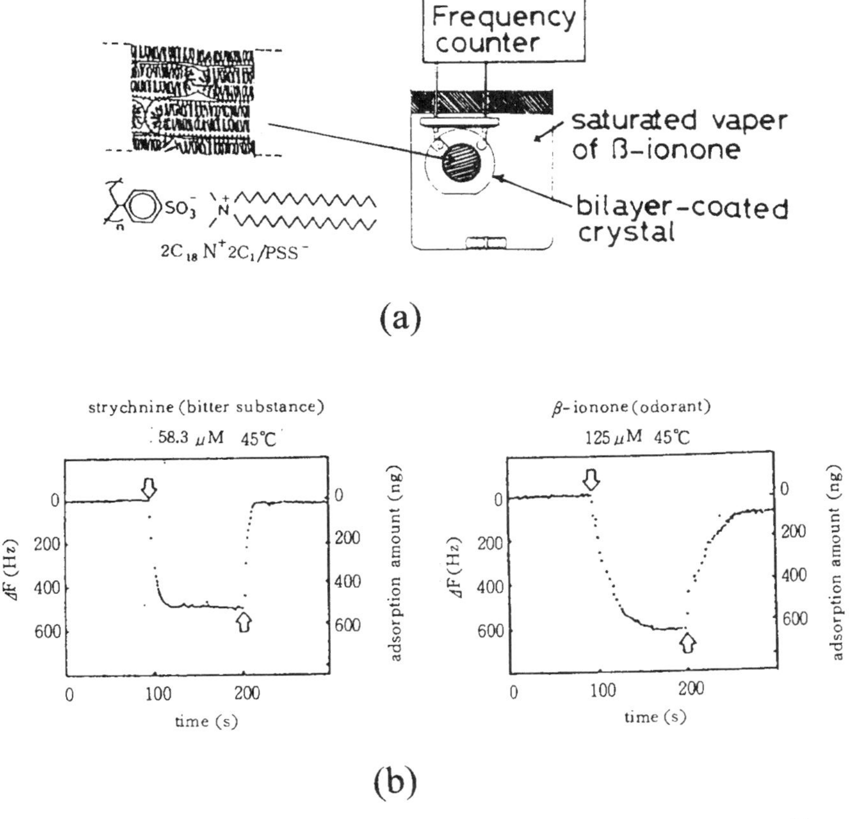

Figure 4.36 Schematic illustrations of a sensor for odor and taste stimuli (a) and its response to a bitter substance (strychnine) and an odorant compound (β-ionone) (b). Figures are drawn with some modifications from the source (Okahata *et al.*, 1987).

mass change of order of nanogram is detectable by this method. Figure 4.36(b) shows the decrement of the proper frequency of the quartz cell when strychnine and β-ionone are added to the solutions as a model compound of bitter substance and odorant, respectively. The response is fast (within 30 seconds) for strychnine and relatively slow (several minutes) for β-ionone. These variations in the response time may be ascribed to the difference in the adsorption part of the compounds in the multiple bilayer membrane film. The strychnine molecules may adsorb only on the surface of the film, but the β-ionone penetrates even inside the film. This bilayer-coated piezoelectric cell works also in the gas phase and thus can detect gaseous odorants.

4.5 Environmental Problems in the Consumer Products and Surfactant Industry

Waste generation is doubtless one of the most serious environmental problems today. The wastes generated from industries and private homes often exert unfavorable influence on natural and social environments. In this section, we will discuss this and other environmental problems in terms of the industries of consumer products and surfactants. We begin with the basic philosophy of the problem and then turn to basic and practical means now used to overcome the problems.

4.5.1 BASIC PHILOSOPHY FOR ENVIRONMENTAL PROBLEMS

a. The Earth as a Thermodynamically Closed System

The earth may be considered a thermodynamically closed system since no material transport is present and only energy transport is possible. If the earth were an isolated system without even energy transportation, thermodynamic equilibrium would be attained sooner or later. On such a thermodynamically equilibrated earth, living things, which must keep a low entropy state in their body to live, would not exist. Fortunately, earth is a closed system with high energy from sunlight coming in and low energy from radiation going out. This energy flow is fundamental for maintaining all the living things on the earth. If the earth were in a steady state, no more living things could be generated than those now present that have been produced by the above-described energy flow. In such a condition, the increase in population of any species must cause the decrease in population of another species. A steady state on the earth including the population of living things guarantees the materials cycle containing the food-chain. If such a steady state were established on the earth, the population of all living things would be kept constant.

To better understand the previous statements, one needs to fully understand why the energy flow from sunshine to radiation is essential to maintain living things. The sunshine has a power spectrum that corresponds to high energy equilibrated with the surface temperature of the sun (~6000 K). One can easily demonstrate this fact by burning a black paper or cloth by focusing the sunshine on it with a lens. One can see that the power spectrum of light equilibrated with a certain temperature shifts to shorter wavelength with increasing temperature, since the light radiated from a heated metal shows blue at high temperatures and red at lower temperatures. The phenomenon displaying a proper power spectrum of light (electromagnetic wave) corresponding to an equilibrated temperature is called *black body radiation.* A photon with short wavelength possesses high energy. So, the sunlight has a high energy corresponding to ~6000 K. The radiation from the earth back into space, on the other hand, possesses much lower energy corresponding to the temperature of the earth (~300 K).

The second law of thermodynamics guarantees that the work (free energy) can be generated like a heat engine when the heat is taken from a high-temperature reservoir and thrown away from a low-temperature one. And the maximum efficiency of the heat engine is $(T1 - T0)/T1$, where T1 and T0 are the absolute temperatures of the reservoirs of higher and lower temperature, respectively. When the heat is taken from the sunshine and thrown away by radiation from the earth, the efficiency of the "engine" is very high, i.e., $(6000 - 300)/6000 = 0.95$. This efficient free energy of sunshine is converted to chemical energy (the synthesis of carbohydrates through carbon dioxide assimilation) by the photosynthesis of plants. And all living things on the earth are maintained by this production of chemicals with high energy (low entropy). Energy is flowing as sunlight → plants → animals → radiant heat, and the materials are cycled between plants and animals. The materials of plants are transferred to the animals as foods, and the materials of animals to the plants as carbon dioxide (from respiration), excrements (fertilizer), and dead bodies. Figure 4.37 shows the most fundamental energy flow and materials cycle in the biological systems on the earth. If the earth were already in the steady state, this energy flow and materials cycle would continue forever.

If any part of the biological cycling system expands abnormally in the steady state of the earth, we must assume that some extra energy and/or materials have been put into the system. The extra energy and materials are fossil fuels and resources. Human beings use a great amount of these fossil fuels and resources—

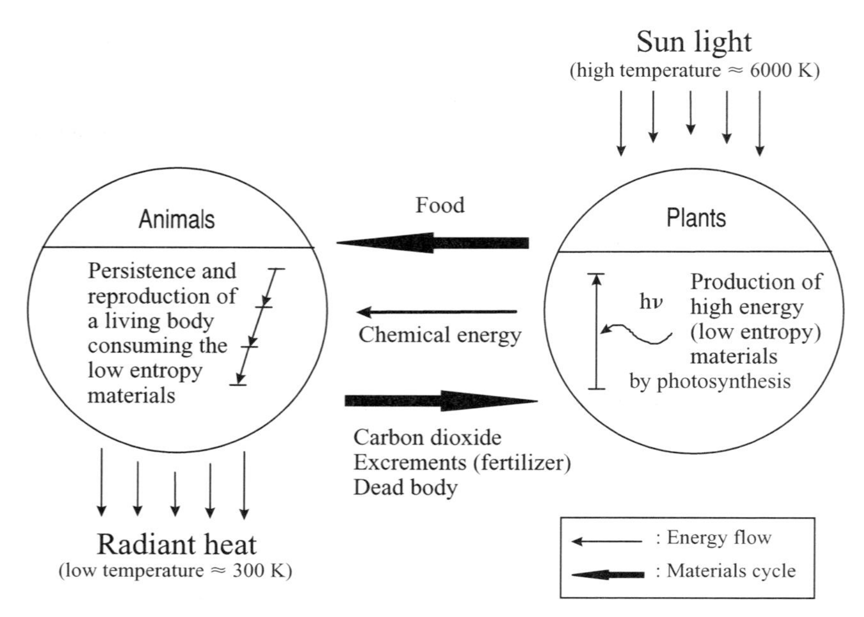

Figure 4.37 The energy flow and the materials cycle between plants and animals on the earth. All living things are generated and maintained by this energy flow.

such as petroleum and coals—in a relatively short period, and the steady flow of energy and the cycle of materials on the earth has been disturbed. This is the environmental problem in the most essential and fundamental sense.

The mass use of fossil fuels has increased the amount of carbon dioxide in the atmosphere upsetting the normal balance: normally, its generation by animal respiration is perfectly offset by its assimilation by plants. In addition, the plastics produced from the fossil resources are not biodegradable and are the chief materials of waste problems. If we develop biodegradable plastics, then the carbon dioxide problem is enhanced. In short, the essence of the problem is too much consumption of materials containing carbon atoms.

The problem snowballs as the population of human beings increases very rapidly as a result of a more affluent life made possible by the use of fossil fuels and resources. Then the food problem occurs. Also, the expansion of the human population causes the decrease in forest areas and the consequent increase of desert areas (recall that an increasing population in any biological species must cause a decrease in the population of another species if the earth is in the steady state). And there is more: Acid rain comes from sulfur compounds present in the fossil fuels, and fluorinated compounds synthesized from fossil resources may destroy the ozone layer. One can now understand that all the present environmental problems originate essentially from the mass and rapid consumption of the fossil fuels and resources. Therefore, the basic means for solving the problems should also come from this origin.

b. Basic Means for Solving Environmental Problems

One can readily understand from the above discussion that the most fundamental way to solve the environmental problems would be to not use fossil fuels and resources. But we cannot return civilization to a pre-Industrial Revolution lifestyle, so this solution is impractical. There are, however, some practical actions that can be implemented. In order of importance, these actions are

1. Do not discard materials to the environment.
2. Use materials repeatedly once produced.
3. Employ materials that add a smaller load to the environment when they inevitably must be discarded.

Let us discuss the concrete means for these actions in the next section.

4.5.2 PRACTICAL MEANS FOR INDIVIDUAL PROBLEMS

a. Do Not Discard Materials to the Environment

This method stops waste materials at their entrance, and thus is the most important and essential one for the environmental problems. Following are several practical means for implementing this method.

Development of products that require smaller amounts of material to perform their functions. We buy and use the functions of products and not the materials themselves. So products that perform the same functions using a smaller amount of the material are desirable. We can reduce the wastes and make the environmental load smaller by using such products. The compact heavy-duty detergent mentioned in Section 4.2.2 is a typical example. Other products in compact sizes—such as fabric softeners, dishwashing detergents, and disposable diapers—are also now available. Further research is expected to provide even more products in this category.

Reducing the packaging materials. Elaborate packaging for department store and/or supermarket goods is neither resource-saving nor waste-reducing and should be simplified. Packaging of the products delivered from a manufacturing plant can also be reduced. For example, detergent cartons can be wrapped simply with a wire without encasing it in another carton. Of course, the compact detergents and other compact products mentioned previously also contribute since they require smaller packages.

Energy saving. Energy saving is also a very important mean to reduce the usage of fossil fuels and the discharge of carbon dioxide. After the oil shortage of the 1970s, most enterprises have endeavored to reduce energy consumption by developing more rational processes for production and distribution, eradicating energy leakage and wastefulness in production plants, utilizing the waste heat generated in burning process of industrial rubbish, and so on. Figure 4.38 shows the energy consumption indexes compared with the gross national product in some major developed countries. Every country has succeeded in saving energy to a considerable extent.

The compact products mentioned previously also contribute to energy savings both in terms of production and the reduction of packaging materials. Such total technologies for energy saving rather than methods to cease energy loss will become much more important in the future.

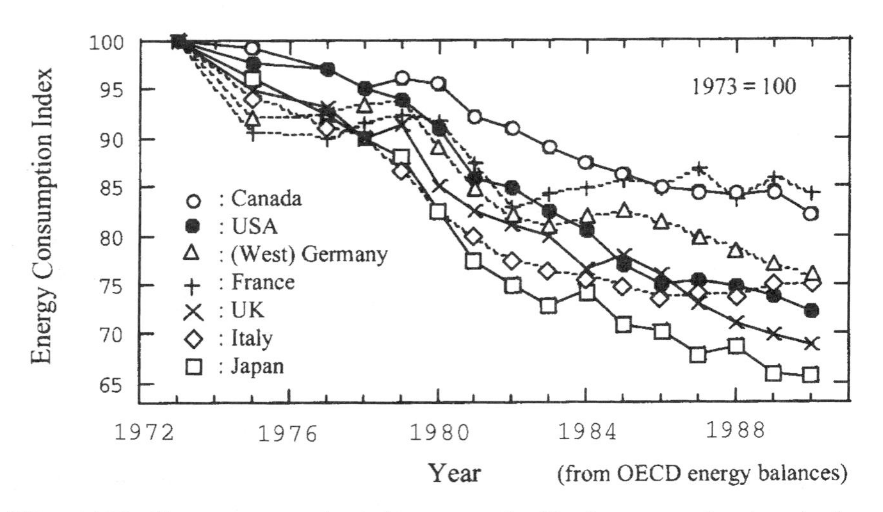

Figure 4.38 Energy consumption indexes normalized by the gross national production in some developed countries (data from OECD).

b. Use Materials Repeatedly Once Produced

Materials are thrown away because they are no longer usable. Accordingly they are not thrown away if they can be used repeatedly. This is a kind of product recycling, and may be the most reasonable recycling plan. Recycling in the literal sense involves recovering, selecting, washing, relabeling, and reusing the used products, say, bottles. On a continuing basis, this process is much more expensive than that to produce new bottles, and may be even more energy-consumptive. Consider all the steps involved: Fuel is consumed and carbon dioxide is discharged by the truck used for recovering, detergent and water are necessary to wash the bottles, that waste water must be treated, and so on. Burning the used bottles and recovering the waste heat is surely more realistic and wise recycling. It is clear that we must rationally consider the total processes for the recycling of various products to find the best solutions for environmental problems.

Recently, standing pouches made of thin polyethylene or paper films have been used to provide powdered or liquid products (such as detergents, softeners, etc.) to fill up the empty, originally purchased bottle. The refillable bottles can be used repeatedly, and both packaging and wastes are reduced. This kind of product favorable for repeated use will be needed more and more in the future.

Deinking agents contribute to the recycling of waste or used papers (newspapers in particular), as mentioned in Section 4.3.1.*c*. In Japan, more than 90% of newspapers are recovered and converted to pulp again by the use of deinking agents. The pulp is reused as newspapers, toilet rolls, and so on. Figure 4.39 shows the rate of paper recycling in several countries.

Utilizing natural resources in the surfactant industry—such as coconut oil, palm oil, and tallow—is also a kind of recycling in the broad sense. Contrary to the irreplaceable fossil resources, these natural resources can be reproduced by plants from the sunshine, water, and carbon dioxide in the atmosphere. So these reproducible resources are in the materials cycle on the earth. It is important for human wealth to thicken the cycle pipe by enhancing the productivity of such resources.

c. Employ Materials That Add a Smaller Load to the Environment

It is inevitable that some products, such as detergents and shampoos, become waste after they are used. Thus, it is important to develop these materials so that they add as small a load as possible to the environment.

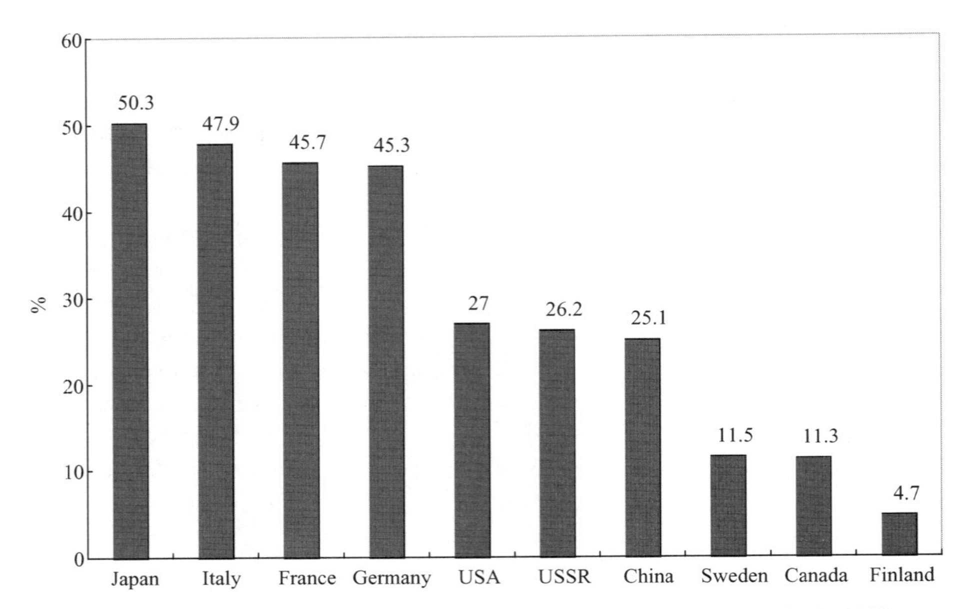

Figure 4.39 Rate of wastepaper recycling in several developed countries in 1989.

Materials compatible with the environment. The raw materials of products must be compatible with the environment so that they will not exert unfavorable influence on the environment. For example, fluorocarbon gas is not compatible with the atmosphere because it destroys the ozone layer. Fluorocarbon gas is now prohibited and aerosol sprays now use hydrocarbons from the fluorocarbons as the propellent gas. Refrigerants used in refrigerators and coolers and the washing solvent for electronics and dry cleaning are still being studied to improve their performances.

Biodegradability and bioaccumulation are two major criteria for compatibility with aqueous systems. Toxicity for small living things, using the water flea as a test standard, is another compatibility check. In biodegradation, organic compounds are assimilated by microorganisms and decomposed finally to carbon dioxide and water. Biodegradation takes place in both water and soil media. The materials that are inevitably discharged to the environment must be easily biodegraded so that the waste materials do not remain in the environment for a long period of time. Some organic compounds are easily biodegraded by microorganisms, but some are not. Branched-chain hydrocarbons are not, so, linear alkylbenzene sulfonate (LAS) is used instead of alkylbenzene sulfonate (ABS). Surfactants synthesized from natural resources (coconut oil, palm oil, tallow, etc.) have a linear hydrocarbon chain, which is easily biodegraded. Any materials that are not easily biodegraded in a water medium must be checked for bioaccumulation in fish, since the bioaccumulated material may be harmful to the fish and/or to whatever might eat the fish (birds, humans, etc.).

Materials free from toxicity. Any raw materials and products formulated therefrom must be free from toxicity. Oral toxicity tests include acute, semiacute, chronic, and gene toxicity. Teratogenicity and cancer generation tests including mutagenic cancer are done for gene toxicity. A gene toxicity test is sometimes carried out through successive generations of animals. Skin and mucous membrane toxicity tests include irritation and allergy tests. An eye irritation test is also performed for special products used in or near the eyes. The toxicity of surfactant is in the order of cationic > anionic > nonionic in both oral and skin toxicity. Every commercial product has, of course, passed through these toxicity tests.

Refreshing Room!

HEART IS MORE IMPORTANT THAN SCIENCE IN DEVELOPMENT

In the development of products, what the consumers want is the most important thing. Science is just a means, not an end. Even those working in the research and development division of a business must have a marketing mind that seeks to know what consumers want and how to satisfy consumers' needs.

Since research workers are also consumers, they can imagine what is useful and convenient in our daily life. Careful observation in their own and their neighbors' lives can reveal hints for new consumer goods. Whether for consumer or industrial use, some insight is necessary to develop the most suitable product. Imagination and insight are essential to understand the wants of customers, and these come from the heart rather than science.

It is often necessary to make a contradiction stand together in developing research. For example, shampoos and dishwashing detergents must have a high-foaming property for washing, but a quick defoaming one for rinsing. Long-lasting sweat-resistant cosmetics must also be able to be cleansed easily. If we have a strong will in heart to make the product complete, we can overcome the contradiction.

Even under such situations as previously described, science is, of course, still necessary and important to develop the products. We cannot make any goods without the aid of scientific knowledge and techniques. Thus, researchers in a private company must improve their scientific skills while keeping the marketing mind in their heart.

Chapter 5 | Surface Active Substances in Biological Systems

5.1 Self-Organization—A Key Phenomenon in Biological Systems

Stratified structure is characteristic of biological systems. An individual body is organized with tissues and organs that are a functionally organized assembly of a number of cells. The cells are composed of organelles (such as nucleus and mitochondria) and the organelles are a molecular assembly of proteins, nucleic acids, polysaccharides, and lipids. Self-organization or self-assembly is the key phenomenon in such stratified structures in biological systems. In each step of the stratified structure, molecules, cells, organs, and tissues are self-organized to form the system of upward structure.

Surface activity contributes to the self-organization of biological systems at the molecular level. As we have mentioned, adsorption and aggregation are two fundamental, characteristic properties of surface active substances (see Section 3.1). Both properties lead to self-organization of biological molecules. For instance, lung surfactants such as phosphatidyl choline work by adsorption onto the lung surface. The adsorption layer of the surfactants depress the surface tension of the lung surface, allowing the lung to expand easily for respiration. Typical examples of the aggregation phenomena in biological systems are biomembranes and liquid crystal formation. Bilayer membranes of surface active substances, phospholipids in particular, show essentially the same structure and functions of biomembranes, and there exist some examples of liquid crystal in biological systems. We will discuss later the contributions of surface activity (adsorption and aggregation) to the self-organization or self-assembly in biological molecules.

Self-organization in biological molecules is due to hydrophobic interactions, hydrogen bonds, ionic interactions, and van der Waals attractions. Among these interactions, hydrophobic interactions and hydrogen bonds are the main factors for the higher-order structures of proteins, nucleic acids, and lipids in aqueous media. They are also the main cause of surface activity in water. Accordingly,

Table 5.1 Self-Organization in Biological Molecules Originated from Surface Activity

	Adsorption	*Aggregation*
Proteins	• Amphiphilic molecules on serum albumin • Membrane proteins on the biomembranes • Substrate (signal) molecules on enzyme (receptor) proteins	• The fluid mosaic model of biomembranes • Channel formation in biomembranes • Multienzyme complexes • Surfactant–protein complexes
Nucleic acids	• Intercalation of aromatic compounds between base residues	• Double helix formation of DNA • DNA–histone complex
Lipids	• Lung surfactants on lung surface • Amphiphilic molecules on serum albumin	• The fluid mosaic model of biomembranes • Bilayer structure of lipid membranes • Liquid crystal formation in tissues and organs

many self-organizing phenomena in biological systems originate from surface activity. These phenomena are summarized in Table 5.1. Some of the important systems will be discussed in the following sections.

5.2 Biomembranes—Self-Assembled System of Lipids and Proteins

5.2.1 STRUCTURE OF BIOMEMBRANES—THE FLUID MOSAIC MODEL

The fundamental structure of biomembranes is the bilayer of lipids (surface active substances). The biomembranes contain proteins in the basic structure of the lipid bilayer membranes. The proteins incorporated in and/or on the lipid membranes are logically called *membrane proteins.* The membrane proteins possess a unique amphiphilic structure in contrast with the water-soluble globular proteins whose surface is surrounded with hydrophilic amino acid residues. The membrane protein molecules have two parts in one molecule: one hydrophilic and the other hydrophobic. The hydrophobic part of the protein molecule is in contact with the lipid bilayer membranes due to the hydrophobic interaction. The origin of this self-organization between lipid bilayers and proteins is essentially the same as that of adsorption of and/or solubilization by surface active substances. Thus, one can regard a membrane protein as a kind of surface active substance.

Singer and Nicolson first proposed the *fluid mosaic model* of biomembranes based on the organized structure of lipid bilayers and membrane proteins mentioned above (Singer and Nicolson, 1972). Figure 5.1 is a schematic representation of this model. The lipid bilayer membranes are in an essentially liquid state (i.e., in a micellar state and not in a gel state; see Section 3.3.4.*a*). Membrane proteins are distributed in a mosaic style on and/or in the liquidlike bilayers. Some proteins sit on the bilayer membranes of lipid with their hydrophobic tip barely attached to the surface of the membrane. Others penetrate inside the bilayers and even pass through the membrane several times (Figure 5.2). Protein molecules can move laterally in the bilayer membranes because of the liquid nature of the membranes.

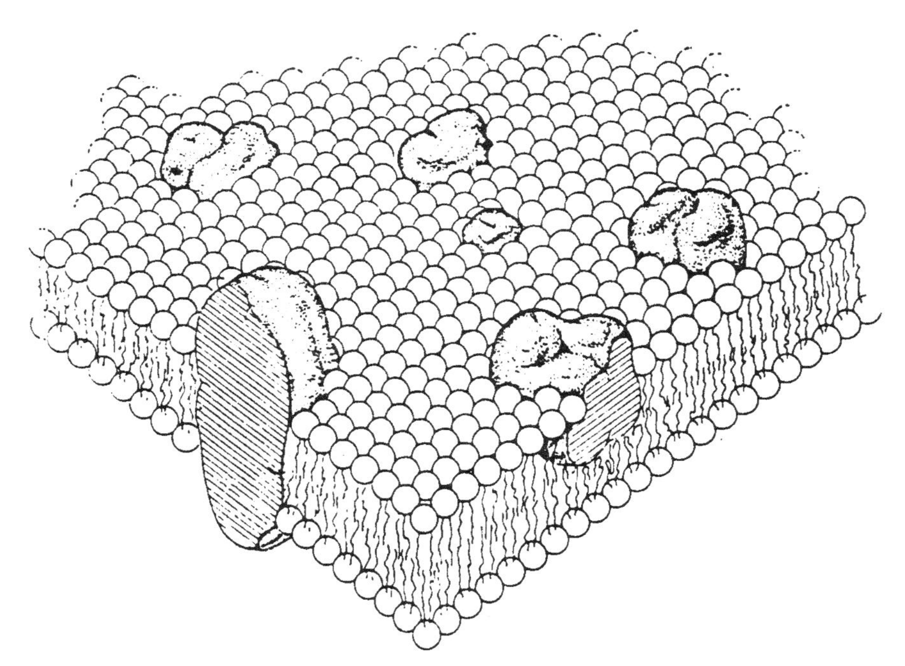

Figure 5.1 The fluid mosaic model of biomembranes. Reprinted with permission from S. J. Singer and G. L. Nicolson, *Science,* **175,** 720 (1972), American Association for the Advancement of Science.

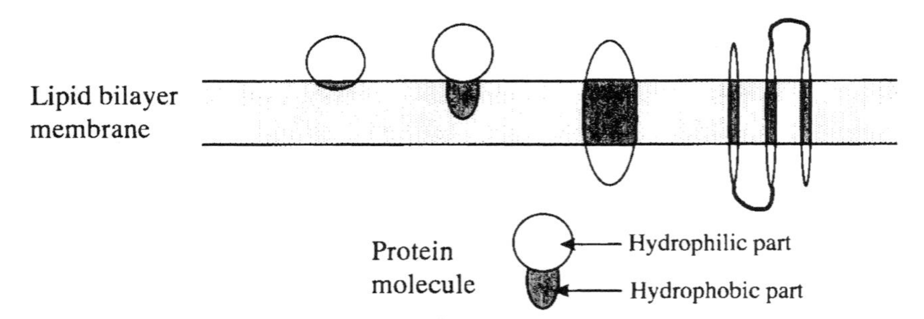

Figure 5.2 Various situations of membrane proteins in the bilayer lipid biomembrane.

The structure of the fluid mosaic model of the biomembrane is established by thermodynamics and not for any apparent biological reasons. The model has been extensively employed since its conception to explain various functions of the biomembranes, some of which will be discussed in the following sections.

5.2.2 FUNCTIONAL ASSEMBLY OF PROTEINS AND OTHER SUBSTANCES IN BIOMEMBRANES

a. Channel Formation by Protein Assembly in Lipid Bilayer Membranes

Many kinds of biologically important substances —such as ions, sugars, amino acids, etc.—are hydrophilic and water-soluble. Thus, they cannot pass through the hydrophobic lipid bilayer membranes. The problem is solved by some smart and sophisticated devices in biological systems that transport these important hydrophilic substances. Channel formation by proteins, peptides, and other substances in the lipid bilayer membranes is one of the most typical examples of such devices.

Ion channels through which certain ions selectively pass play one of the most important roles in taste reception on an animal tongue (Gilbertson, 1993). Five different tastes (sweet, bitter, salty, sour, and umami) are sensed by taste cells on a tongue through two different mechanisms. The first is the receptor mechanism: The taste substances bind to a receptor protein and the stimulus is transferred to the inside of the cell. The second is the ion-channel mechanism: The taste stimuli open the K^+ or Na^+ ion channel and change the membrane potential, which is maintained by the concentration difference of K^+ and Na^+ ions between the inside and the outside of the cell membranes. The ion channels consist of some proteins, but the molecular structure of the channels is still unclear. Much effort is being applied to elucidate the molecular mechanisms of the ion permeation (Findlay and Marsh, 1995). Similar ion channels work in nerve cells to fire and transfer the nerve excitation through the membrane potential change of the cell.

In addition to physiological ion channels, there are several channel-forming antibiotics—nystatin, amphotericin B, and gramicidin A, are typical. Some of them are peptides and some are not. Nystatin and amphotericin B are polyene molecules very similar to each other, and are assumed to form an ion channel as shown in Figure 5.3 (Kleinberg and Finkelstein, 1984). Ten molecules of the antibiotics form an ion channel that provides a hydrophilic pore inside the channel and a completely hydrophobic exterior surface that is in contact with the lipid bilayer membrane. The inner diameter of the channel pore is about 0.8 nm, reflected by the size dependence of the permeable molecules. The antibiotic activity of these antibiotics is, of course, believed to be due to the ion-channel formation in biological membranes of bacteria, which results in a fatal imbalance of ions between the inside and the outside of the bacterial cells.

Artificial channel-forming materials also have been studied extensively. Mimicking the channel-forming antibiotics, some of the artificial materials are peptides and some are not. A self-assembled cyclic peptide of eight amino acids has been synthesized and tested as a channel-forming compound for permeation of hydrophilic molecules (Ghadiri *et al.*, 1994), as shown in Figure 5.4. The molecules are stacked in an antiparallel arrangement and are connected firmly by many intermolecular hydrogen bonds. Their hydrophobic side chains are in contact with a lipid bilayer membrane. This channel-forming peptide works even

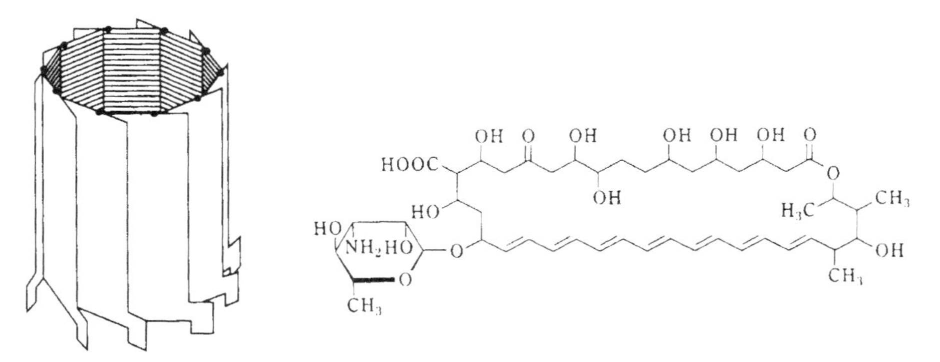

Figure 5.3 Schematic model of an ion channel formed by polyene antibiotics (nystatin) in the lipid bilayer membrane. M. E. Kleinberg and A. Finkelstein, *J. Membrane Biol.*, **80**, 257 (1984).

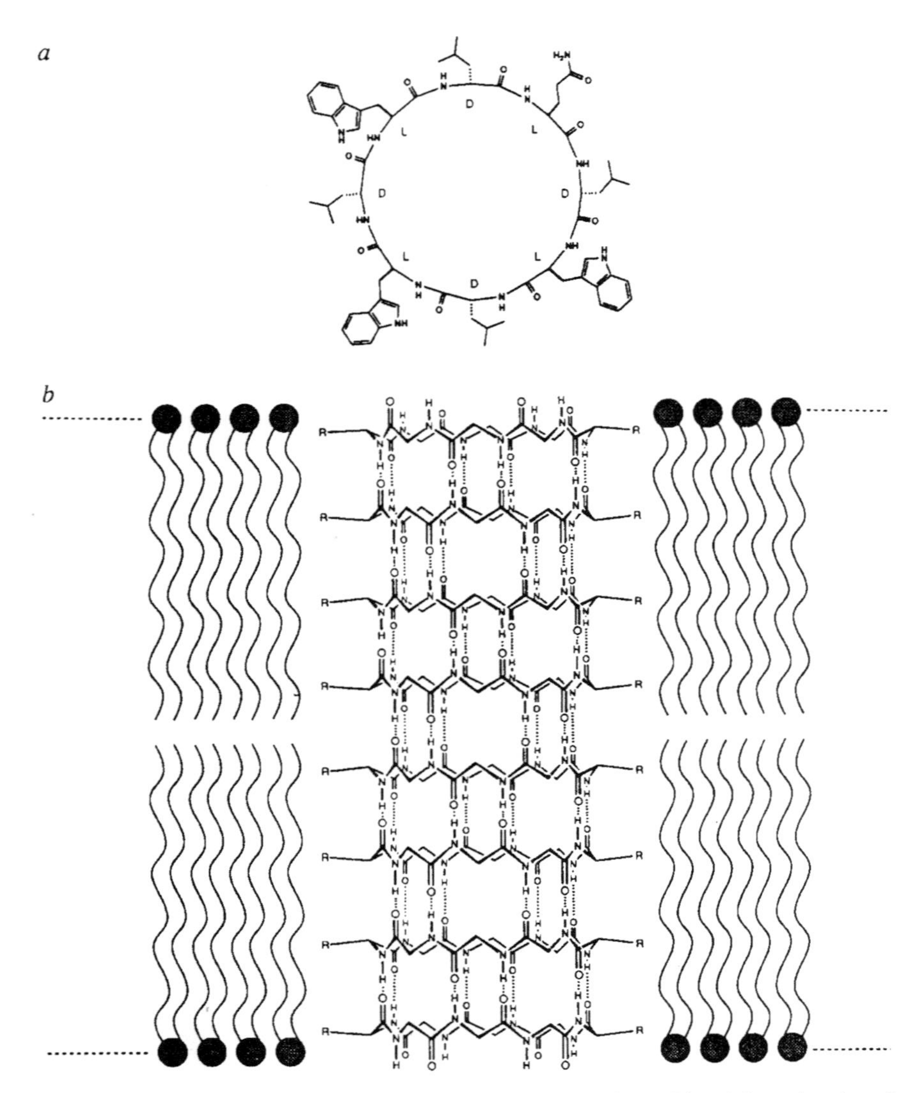

Figure 5.4 Schematic illustration of an artificial channel formed by eight molecules of a synthetic cyclic peptide (Ghadiri *et al.*, 1994). Reprinted with permission from *Nature,* **369,** 301 (1994), Macmillan Magazines Limited.

better than gramicidin A or amphotericin B for proton permeation through phosphatidylcholine liposomes. Artificial channel-forming materials other than peptides also have been reported, such as amphiphilic compounds (Kobuke *et al.*, 1992) and crown-ether-type ring compounds (Fyles *et al.*, 1993). These also work as an ion channel by self-assembling in the lipid bilayers.

b. Multienzyme Systems

Successive chemical reactions take place quite frequently in biological systems; biosynthesis of fatty acids is a typical example. In such reactions, the enzyme for each reaction must be arranged in the proper sequence, and a self-assembled multienzyme system (or complex) provides the required sequence (e.g., Mahler and Cordes, 1971). In another type of multienzyme system, not only does each enzyme in the complex catalyze a proper reaction, but the multienzyme system unit also works as a completely new enzyme that catalyzes a different reaction. The pyruvate dehydrogenase system is a typical example of this second type of multienzyme complex (e.g., Mazur and Harrow, 1971). In all cases of multienzyme systems, the self-assembling property of the enzyme molecules is the essential key to their activity.

5.2.3 LIQUID CRYSTALS IN BIOLOGICAL SYSTEMS

Many liquid crystal structures are known in biological systems (Brown and Wolken, 1979). Liquid-crystal-forming materials are lipids, proteins, carbohydrates, and other biological compounds. Some liquid crystal formations seem independent of surface activity and thus are not discussed here (e.g., muscle and connective tissue, see Hawkins and April, 1983). But others are surely formed due to surface activity or by surface active substances, some in lamellar structures and some in cubic ones.

a. Lamellar Liquid Crystals in Biological Systems

The photosynthetic systems of plants and the visual systems of animals are two major systems that show the photoresponsive function. The photosynthesis of carbohydrates through assimilation of carbon dioxide by plants is the origin of the successive food chain that maintains life on the earth (see Section 4.5.1.*a*). The visual system is, of course, a very important organ for animals to see the environment and respond to it. Very interestingly, both systems contain beautiful lamellar structures in their photoresponsive devices (Brown and Wolken, 1979).

The photoresponsive pigments—chloroplast in the thylakoid membrane of plants and rhodopsin in outer segment of the retinal rod and cone cells in animals—are stacked in the lamellar structure. The pigments located in these lamellar structure of membranes are very efficient in catching the sunlight because of the extremely large surface area of the structure.

Lamellar liquid crystals are also found in the stratum corneum (the top layer) of the skin. The stratum corneum is a thin (~30 μm) layer, but plays an important role as the barrier to water loss from a living body as well as to the invasion of materials from external environments. Figure 5.5 shows a transmission electron microscope photograph of a stratum corneum; one can see the lamellar structure in the intercellular spaces. This lamellar structure consists of bilayer membranes of lipids such as ceramides, fatty acids, cholesterol, and so on. These lipids are different from sebaceous ones because of their origin: Sebaceous lipids are secreted from the sebaceous gland, but the stratum corneum lipids originate from cell membranes of the epidermal cell. Because hair and skin are related and the

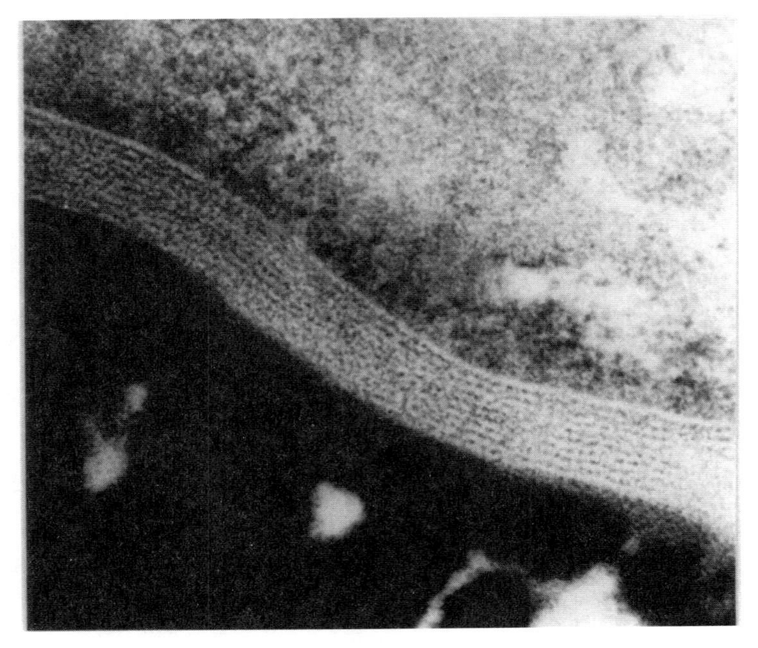

Figure 5.5 A transmission electron microscope photograph of a stratum corneum. A beautiful lamellar structure can be seen in the intercellular spaces. Reprinted by courtesy of Dr. T. Suzuki, Kao Corporation.

cell membrane complex (CMC) of the hair is also of lamellar structures. The CMC is located in between the cortical cells of hair and is assumed to cement the cells (Robbins, 1994).

b. Cubic Liquid Crystals in Biological Systems

Examples of cubic liquid crystalline phases found in biological systems are etioplasts and endoplasmic reticulum (Lindblom and Rilfors, 1989). The etioplasts are precursors of chloroplasts, formed in plant leaves grown in the dark. They turn to chloroplasts when exposed to light. The etioplasts consist of a tetrapodal unit structure of tubules that resembles the structure of a bicontinuous cubic phase. Endoplasmic reticulum is also constructed of tubular networks having a cubic symmetry, which again is very similar to the bicontinuous cubic phase. Although the cubic liquid crystalline phases of these organs must play a role in maintaining the biological systems, the significance of the cubic structure is not yet clear.

c. Moisturizing of Skin by Liquid Crystals

Several bilayer membranes are present between cells of the stratum corneum to form a lamellar structure, as mentioned above (Figure 5.5). This lamellar structure plays an important role in moisturizing the skin (Imokawa *et al.*, 1986, 1991; Grubauer *et al.*, 1989). Skin becomes rough and dry when treated with some organic solvents, like an acetone/ether mixture. The skin materials extracted by such a treatment have been analyzed and found to be mainly ceramide (a sphingolipid) and other lipids such as cholesterol and fatty acids. Thus, these lipids would moisturize the skin if reapplied. These lipid mixtures have been shown to form a lamellar liquid crystal in water medium, which is important for moisturizing.

Natural ceramide is located mainly in animal brain and is very expensive to extract and use in skin creams. In addition, some people object to the use of animal products in cosmetics. Thus, an artificial ceramide has been synthesized and formulated in cosmetic skin creams together with the other lipids (cholesterol and fatty acids). The skin creams containing the synthetic ceramide were very effective for skin moisturizing (Suzuki *et al.*, 1993). The molecular structures of both natural and synthetic ceramides are shown in Figure 5.6.

Water-soluble compounds of low molecular weight—such as glycerin and hydrophilic amino acids—have long been known to be useful for moisturizing the skin; these compounds are called the *natural moisturizing factor* or NMF. We now understand, however, that only NMF is not enough. The lipids that form

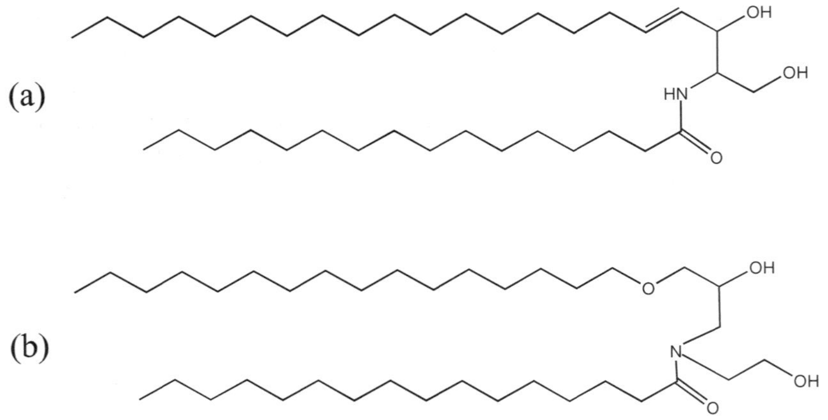

(a)

(b)

Figure 5.6 Chemical structures of natural (a) and synthetic (b) ceramides (Suzuki *et al.*, 1993).

lamellar liquid crystal structures are also needed for the most effective skin moisturizing.

d. Percutaneous Absorption of Medicines by Liquid Crystals

It is widely known that the development of a new drug is a very expensive project. So the research work to give new functions to known drugs has attracted much attention. Percutaneous absorption is one of the most important functions so researched. However, if the chemical structure of a drug is modified to give it the percutaneous absorption ability, the modified compound is already a new medicine. Thus, a percutaneous absorption enhancer that does not show any medical efficacies itself may be the preferred route.

Certain surfactants that are not water-soluble and readily form liquid crystals have been found to be effective as percutaneous absorption enhancers: isodiglycerin dialkyl ether (IDGE: Kamiya *et al.*, 1987) and α-monoisostearyl glyceryl ether (GE: Kawamata *et al.*, 1987). As already mentioned in Section 4.4.2.*a*, the GE forms an emulsion stabilized by a liquid crystal. Indomethacin® (an anti-inflammatory medicine) has been dissolved in the above emulsion and applied to a rabbit skin. The concentration of Indomethacin in the rabbit blood was determined, and the activity of both the GE and the IDGE as an accelerator of percutaneous absorption was tested. The emulsion with the GE was as much as ten times more effective compared with commercial Indomethacin ointments; that with the IDGE was even more so. Figure 5.7 shows the effect of the IDGE on the percutaneous absorption of Indomethacin. The best result on the graph is as effective as AZONE®, which is well known as the most powerful accelerator. It is not clear at present why and how the liquid crystals of GE and IDGE enhance the percutaneous absorption of the drug. It may be that the lamellar liquid crystal present in the stratum corneum are modified or perturbed, thus positively affecting the pathway for drugs.

e. Diseases Caused by Liquid Crystal Formation

Liquid crystal formation in biological systems is sometimes harmful to the living body itself. The liquid crystalline phase of some lipid mixtures can

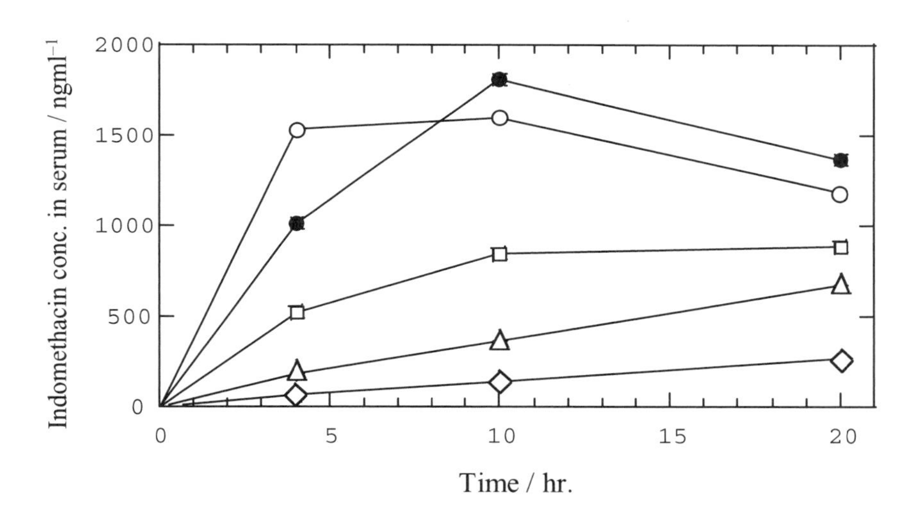

R_1—O—CH_2
|
CH—O—C(H_2)—C(H)—CH_2
| | |
R_2—O—CH_2 OH   OH

R_1=C_8H_{17}, R_2=CH_3 : ◇
R_1=$C_{10}H_{21}$, R_2=CH_3: △
R_1=$C_{12}H_{25}$, R_2=CH_3 : □
R_1=$C_{14}H_{29}$, R_2=CH_3 : ○
AZONE :●

IDGE derivatives:
1-O-R_1-3-O-R_2-2-O-2',3'-dihydroxypropylglycerol

Figure 5.7 Effect of IDGE derivatives on the percutaneous absorption of Indomethacin (Kamiya *et al.*, 1987).

precipitate out into some viscera and disturb their normal function. There have been many such kinds of disease discovered so far (Small, 1977; Brown and Wolken, 1979). Cholesterol ester storage disease in the liver, arteriosclerosis, Tangier disease—these all result from the deposition or accumulation of some liquid crystal containing mainly the cholesterol esters. Gallstone is a crystal of cholesterol monohydrate precipitated out of the liquid crystalline phase in human bile fluids.

Sometimes certain lipids accumulate due to the lack of the enzymes that digest or transform the lipids. Lecithin and sphingomyelin (Neimann-Pick disease), glucosyl ceramide (Gaucher's disease), galactosyl ceramide (Krabbe's disease) and ceramide trihexoside (Fabry's disease) are known examples. The accumulated lipids form liquid crystalline phases together with other lipids in various tissues of the liver, kidney, spleen, and so on.

5.3 Surface Active Compounds Working in Biological Systems

5.3.1 EMULSIFICATION AND SOLUBILIZATION BY NATURAL AMPHIPHILES IN RELATION TO PHYSIOLOGICAL FUNCTIONS

a. Digestion of Fatty Acid Triglycerides

Hydrophobic and water-insoluble compounds are frequently important in physiological functions. For example, the digestion of fatty acid triglycerides is one of the most essential processes in biological systems to maintain life. The triglycerides must be hydrolyzed by certain lipases to be digested and then the hydrolyzed products—such as fatty acids and monoglycerides—are absorbed into the cells of the small intestine. Both the hydrolysis reaction and the absorption onto the intestinal surface are very dependent on the solution state of the target materials (oils).

Emulsification and solubilization of the triglycerides by bile salts play the essential role in both actions (Hadley, 1985). The hydrolysis reaction rate is, of course, affected by the interfacial area between the oil (triglycerides) and the water phase because the lipase works at the surface of the oils. Bile salts emulsify the triglycerides to make small droplets, thus enhancing the enzymatic reaction rate of lipase for triglyceride digestion. The hydrolytic products of the triglycerides (fatty acids and monoglycerides) are solubilized into the micelles of the bile salts and then transported from the intestinal lumen to microvilli. There, only the hydrolytic products are absorbed at the surface of the microvilli, not the bile salts. The bile salts return to the intestinal lumen to solubilize more digested triglycerides. The bile salt micelles thus work as a transport vehicle for hydrolyzed triglycerides.

b. Hemolysis

Surface active compounds sometimes cause the hemolysis of red blood cells. Soaps and synthetic surfactants in particular, when used at high concentrations, may solubilize and break down the erythrocyte membrane causing hemoglobin to come out of the erythrocytes. But the hemolysis takes place even at concentrations of surface active compounds lower than CMC: The surface active compounds adsorb and perturb the bilayer membrane structure of erythrocyte and cause the hemoglobin molecules to pass through the membranes. The detailed mechanism of this membrane perturbation has not yet been made clear.

Some natural compounds such as lyso-phospholipids and melittin also show the hemolytic activity. The lyso-lipids are single-chain surfactants and their hemolytic activity is straightforward. Melittin is the venom of a bee and acts strongly on membranes in several ways (Dempsey, 1990). A small protein having 26 amino acid residues, melittin causes the hemolysis and even solubilization of erythrocyte membranes at high concentrations. Other actions of this small protein on the membrane properties are voltage-gated channel formation and membrane fusion. The melittin molecule is interestingly a kind of amphiphile that consists of both hydrophobic and hydrophilic parts. This amphiphilic nature of the melittin seems to play an important role in its actions on the membranes, but the detailed mechanisms are not yet clear.

5.3.2 INTERACTION OF SURFACE ACTIVE COMPOUNDS WITH PROTEINS AND NUCLEIC ACIDS

a. Binding of Fatty Acids and Drugs to Serum Albumin

Serum albumin—the most abundant protein in blood components—works as a transport protein between tissues and organs of fatty acids as well as other hydrophobic substances (Brown and Shockley, 1982 as a review). Serum albumin has many binding sites for fatty acids, about six of which have high affinity for fatty acids and more than 20 of which have low affinity. The association constant of fatty acids to the binding site of high affinity is in the order of 10^7–10^9. The binding constant increases with increasing chain length of the fatty acids, and the main driving force is assumed to be hydrophobic interaction.

Since the driving force of the binding is hydrophobic interaction, serum albumin also has the ability to bind other hydrophobic compounds such as steroids, tryptophan, surfactants, etc. Interaction with surfactants will be discussed in the next subsection; here we focus on the binding of drugs. Transportation of drugs into tissues is the main therapeutic action of the albumin. In some cases, however, the binding of the drugs themselves is important to protect against a toxic effect of the drugs. Primary binding sites for drugs are known to be different from those for fatty acids, so they are not competitive with each other. This situation is favorable for both drugs and fatty acids, since they can perform their therapeutic and physiological functions independently.

b. Interaction of Surfactants with Proteins and Nucleic Acids

Surfactant molecules also bind to many kinds of proteins. But the binding of surfactant to proteins shows different features from that of fatty acids to serum albumin. For example, the surfactant binding frequently causes protein denaturation, which is unlikely in fatty acid binding. Figure 5.8 shows a typical binding isotherm of surfactant (sodium dodecyl sulfate) to a protein (lysozyme) (Jones, 1996). The steep increase in the binding amount at lower surfactant concentrations is due to the electrostatic attractive interaction and/or hydrophobic interaction between the hydrocarbon chain of the surfactant and the hydrophobic pockets in the protein molecules. The binding in this region is commonly a Langmuir type. After passing through an almost plateau region, the binding isotherm exhibits again a rapid increase. This second steep increase in the binding is a cooperative one, and the protein molecules are denatured in this step. The denaturation of the protein exposes more hydrophobic residues that are initially buried inside the protein molecule to the solution side. More surfactant molecules bind to the newly exposed hydrophobic residues and denature the protein molecules further. This is why the cooperative and increased and binding takes place accompanying the protein denaturation. Not all surfactants denature the proteins; those that do not are useful to extract membrane proteins from biomembranes and will be discussed in Section 5.3.2.*d*. For more information on the extensively studied surfactant-protein interactions, see, for example, Steinhardt and Reynolds, 1969 and Tanford, 1980c.

Studies on the interaction of surfactants with nucleic acids are very few because the interactions between them are very few. The possible effects of surfactants on the thermal denaturation behavior of the double helix of DNA were examined by Tsujii and Tokiwa (1977). It was found that ionic surfactants do

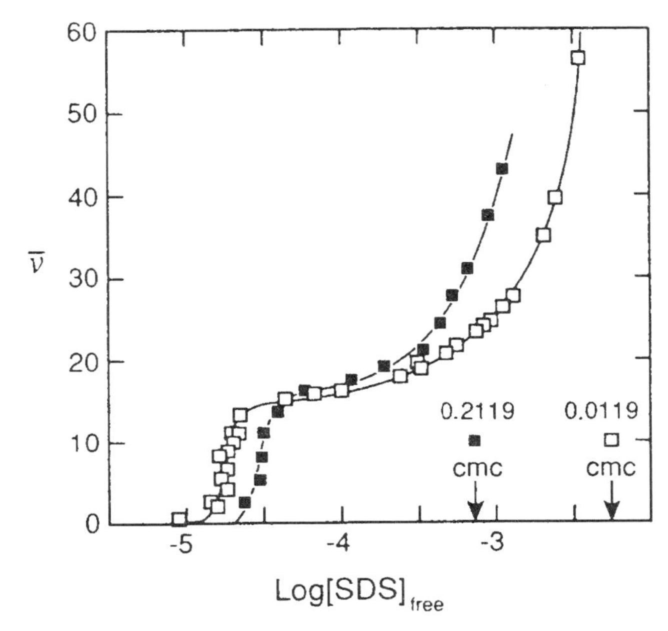

Figure 5.8 Binding isotherm of sodium dodecyl sulfate to lysozyme (Jones, 1996). A typical curve of surfactant binding to a protein. Reprinted from "Surface Activity of Proteins", p. 245 by courtesy of Marcel Dekker, Inc.

increase the melting temperature of the DNA helices, but the effect is just due to the increase of ionic strength. No specific interaction between any kind of surfactants and DNA was observed.

c. SDS-Polyacrylamide Gel Electrophoresis

SDS (sodium dodecyl sulfate)-polyacrylamide gel electrophoresis is a powerful tool and is extensively employed in the field of biochemistry to analyze the molecular weight of proteins. In this technique, proteins are denatured and their S–S bonds, if any, are cleaved. SDS molecules bind to the protein polypeptide chain to form a complex between them. The saturated binding amount of SDS to the protein polypeptide is almost the same irrespective of the protein species (Reynolds and Tanford, 1970; Takagi *et al.*, 1975). This binding causes the proteins to lose their proper character, and they behave like simple polyelectrolytes. Polyelectrolyte molecules are separated by their molecular weight when they move inside the chain networks of the polyacrylamide gel. Small molecules can easily migrate in the gel networks due to their smaller resistance, and thus they move faster in the electrophoretic process. Of course, the protein molecules are actually separated by the hydrodynamic radius (Stokes radius) rather than the molecular weight itself. Thus, the noncharacteristic behavior of the protein polypeptide–SDS complexes is essential for estimating the molecular weight by this technique: The complex formation of SDS with protein polypeptides is the key in the SDS-polyacrylamide gel electrophoresis technique.

d. Extraction of Membrane Proteins by Surfactants

Membrane proteins have hydrophobic portion(s) in their molecules and are generally water-insoluble. So the solubilization of the proteins by surfactants is an important technique for extracting and separating the proteins from biomembranes. The surfactants employed in this extraction method must satisfy the following conditions.

1. Their solubilizing power must be strong enough to extract even integrated proteins.
2. They must be mild enough to keep the proteins' structures and activities intact.
3. They must be easily removed for purification by means of, say, dialysis.
4. They must be inexpensive enough to be used in the common laboratories of universities.

Alkylglucoside, bile salts, Triton X-100 (polyoxyethylene ($\overline{p}$ = 10) octylphenyl ether), etc. are frequently employed as suitable surfactants for the above conditions.

Extracting conditions are, of course, important to obtain the target protein. It is commonly believed that a surfactant concentration above CMC is required to solubilize most membrane proteins. The micelle formation is indeed necessary to solubilize small organic molecules, but this may not be necessarily the case in the extraction of proteins the same size as the surfactant micelles. Figure 5.9 shows a schematic illustration of a solubilized surfactant–protein complex. In this complex, the surfactant molecules adsorb on the hydrophobic part of the protein surface rather than solubilize it inside their micelles. Adsorption of the surfactant molecules onto the polymer chains takes place at a much lower concentration of CMC. This may also be the case in the solubilizing process of the membrane proteins.

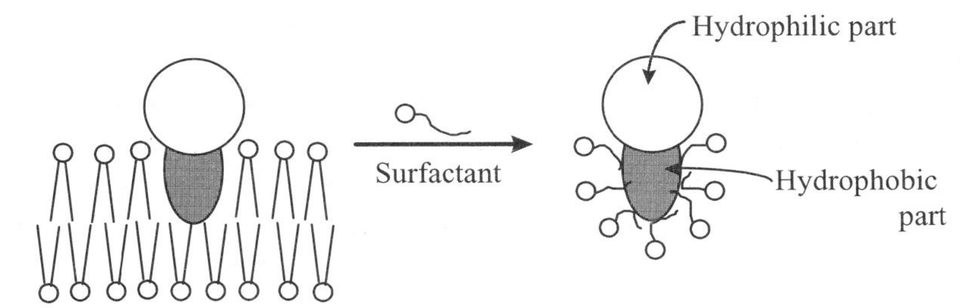

Figure 5.9 Schematic illustration of the solubilization mechanism of membrane proteins by a surfactant. Surfactant molecules adsorb on the hydrophobic part of the protein molecule and solubilize it into water.

Extraction of membrane proteins by liposomes containing the boundary lipids (particularly artificial DDPC, see Section 3.3.4.*c*) seems to be a much better method than the surfactant-extraction one because the native structure and orientation of the extracted proteins are maintained in the liposomal membranes. But a direct comparison of both methods for the extraction of the same membrane protein has never been made. Thus, research to find the best method for each membrane protein is still expected.

References

Adamson, A. W,. *Physical Chemistry of Surfaces,* 4th ed., p. 51, John Wiley, New York, 1982a.

Adamson, A. W., *Physical Chemistry of Surfaces,* 5th ed., p. 387, John Wiley, New York, 1990a.

Adamson, A. W,. *Physical Chemistry of Surfaces,* 5th ed., chs. 5 & 6, John Wiley, New York, 1990b.

Adamson, A. W., *Physical Chemistry of Surfaces,* 4th ed., p. 125, John Wiley, New York, 1982b.

Ahmad, S. I., Shinoda, K., and Friberg, S., *J. Colloid Interface Sci.* **47:** 32–37 (1974).

Alder, B. J. and Wainright, T. E., *Phys. Rev.* **127:** 359–361 (1962).

Alder, B. J., Hoover, H. G., and Young, D. A., *J. Chem. Phys.* **49:** 3688–3696 (1968).

Bader, H., Dorn, K., Hupfer, B., and Ringsdorf, H., *Adv. Polm. Sci.* **64:** 1–62 (1985).

Bancroft, W. D., *J. Phys. Chem.* **17:** 501–519 (1913).

Bancroft, W. D., *J. Phys. Chem.* **19:** 275–309 (1915).

Bangham, A. D., Hill, M. W., and Miller, N. G., *Methods in Membrane Biology,* Vol. 1, ch. 1, E.D. Korn, Ed., Plenum Press, New York (1974).

Bangham, A. D., Standish, M. M., and Watkins, J. C., *J. Mol. Biol.* **13:** 238–252 (1965).

Batzri, S. and Korn, E. D., *Biochim. Biophys. Acta* **298:** 1015–1019 (1973).

Becher, P., *Emulsions: Theory and Practice,* Reinhold, New York, 1957.

Becher, P,. *Encyclopedia of Emulsion Technology, Vol. 1: Basic Theory,* P. Becher, Ed., Marcel Dekker, New York, 1983.

Becher, P,. *Encyclopedia of Emulsion Technology, Vol. 2: Applications,* P. Becher, Ed., Marcel Dekker, New York, 1985.

Becher, P,. *Encyclopedia of Emulsion Technology, Vol. 3: Basic Theory/Measurement/Applications,* P. Becher, Ed., Marcel Dekker, New York, 1988.

Beck, J. S., Vartuli, J. C., Roth, W. J., Leonowicz, M. E., Kresge, C. T., Schmitt, K. D., Chu, C. T-W., Olson, D. H., Sheppard, E. W., McCullen, S. B., Higgins, J. B., and Schlenker, J. L., *J. Am. Chem. Soc.* **114:** 10834–10843 (1992).

Blok, M. C., de Gier, J., and van Deenen, L. L. M., *Biochim. Biophys. Acta* **367:** 202–209 (1974a).

Blok, M. C., de Gier, J., and van Deenen, L. L. M., *Biochim. Biophys. Acta* **367:** 210–224 (1974b).

Bondi, A., *Chem. Rev.* **52:** 417–458 (1953).

Bourrel, M. and Schechter, R. S., *Microemulsions and Related Systems—Surfactant Science Series 30,* Marcel Dekker, New York, 1988.

Brown, J. R. and Shockley, P., *Lipid-Protein Interactions,* Vol. 1, ch. 2, P. C. Jost and O. H. Griffith, Eds., John Wiley, New York, 1982.

Brown, G. H. and Wolken, J. J., *Liquid Crystals and Biological Structures,* Academic Press, New York, 1979.

Charles, S. W. and Popplewell, J., *IEEE Trans. Magn.* **Mag-16:** 172–177 (1980).

Chen, S. H. and Rajagopalan, R., *Micellar Solutions and Microemulsions,* Springer-Verlag, Heidelberg, 1990.

Claesson, P. M., Kjellander, R., Stenius, P., and Christenson, H. K., *J. Chem. Soc., Faraday Trans.* 1, **82:** 2735–2746 (1986).

Davies, J. T. and Rideal, E. K., *Interfacial Phenomena,* ch. 1, Academic Press, New York, 1963.

de Feijter, J. A., *Thin Liquid Films—Surfactant Science Series 29,* ch. 1, I. B.Ivanov, Ed., Marcel Dekker, New York, 1988.

Deamer, D. and Bangham, A. D., *Biochim. Biophys. Acta* **443:** 629–634 (1976).

Deguchi, K. and Mino, J., *J. Colloid Interface Sci.* **65:** 155–161 (1978).

Dempsey, C. E., *Biochim. Biophys. Acta* **1031:** 143–161 (1990).

Ekwall, P., *Advances in Liquid Crystals,* Vol. 1, p. 127, G. H. Brown, Ed., Academic Press, New York, 1975.

Ekwall, P., Mandell, L., and Fontell K., *J. Colloid Interface Sci.* **33:** 215–235 (1970).

Feinstein, M. E. and Rosano, H. L., *J. Phys. Chem.* **73:** 601–607 (1969).

Fendler, J. H. and Tundo, P,. *Acc. Chem. Res.* **17:** 3–8 (1984).

Findlay, J. B. C. and Marsh, D., *Ion Channels—A Practical Approach,* ch. 9, R. H. Ashley, Ed., Oxford University Press, Oxford, 1995.

Frank, H. S. and Evans, M. W., *J. Chem. Phys.* **13:** 507–532 (1945).

Franses, E. I., Talmon, Y., Scriven, L. E., Davis, H. T., and Miller, W. G., *J. Colloid Interface Sci.* **86:** 449–467 (1982).

Friberg, S. E. and Bothorel, P., *Microemulsions: Structure and Dynamics,* S. E. Friberg and P. Bothorel, Eds., CRC Press, Boca Raton, FL, 1987.

Friedel, J,. *Dislocations,* ch. 14, Pergamon Press, New York, 1964.

Fujihira, M., Nishiyama, K., and Yamada, H., *Thin Solid Films* **132:** 77–82 (1985).

Fukuda, K., Shibasaki, Y., and Nakahara, H., *Thin Solid Films* **99:** 87–94 (1983).

Fukuda, K., Shibasaki, Y., and Nakahara, H., *Thin Solid Films,* **133:** 39–49 (1985).

Funasaki, N., *Mixed Surfactant Systems—Surfactant Science Series 46,* ch. 5, K. Ogino and M. Abe, Ed., Marcel Dekker, New York, 1993.

Funasaki, N. and Hada, S., *J. Phys. Chem.* **84:** 736–744; 1868–1869 (1980).

Fyles, T. M., James, T. D., and Kaye, K. C., *J. Am. Chem. Soc.* **115:** 12315–12321 (1993).

Gabriel, N. E. and Roberts, M. F., *Biochemistry* **23:** 4011–4015. (1984).

Gabriel, N. E. and Roberts, M. F,. *Biochemistry* **25:** 2812–2821 (1986).

Gee, M. L., McGuiggan, P. M., Israelachvili, J. N., and Homola, A. M., *J. Chem. Phys.* **93:** 1895–1906 (1990).

Ghadiri, M. R., Granja, J. R., and Buehler, L. K., *Nature* **369:** 301–304 (1994).

Gilbertson, T. A., *Current Opinion in Neurobiology* **3:** 532–539 (1993).

Gillberg, G., Lehtinen, H., and Friberg, S. E., *J. Colloid Interface Sci.* **33:** 40–53 (1970).

Goddard, E. D. *J. Soc. Cosmet. Chem.* **41:** 23–49 (1990).

Goddard, E. D. and Ananthapadmanabhan, K. P., *Interactions of Surfactants with Polymers and Proteins,* E. D. Goddard and K. P. Ananthapadmanabhan, Eds., CRC Press, Boca Raton, FL, 1993.

Goddard, E. D. and Benson, G. C., *Can. J. Chem.* **35:** 986–991 (1957).

Granato, A. V., Lucke, K., Schlipf, J., and Teutonico, L. J., *J. Appl. Phys.* **35:** 2732–2745 (1964).

Griffin, W. C., *J. Soc. Cosmet. Chem.* **5:** 249–256 (1954).

Grubauer, G., Fingold, K. R., Harris, R. M., and Elias, P. M., *J. Lipid Res.* **30:** 89–96 (1989).

Guering, P. and Lindman, B., *Langmuir* **1:** 464–468 (1985).

Hachisu, S. and Kobayashi, Y., *J. Colloid Interface Sci.* **46:** 470–476 (1974).

Hadley, N. F,. *The Adaptive Role of Lipids in Biological Systems,* pp. 302–308, John Wiley, New York, 1985.

Hawkins, R. J. and April, E. W,. *Advances in Liquid Crystals,* Vol. 6, pp. 243–264, G. H. Brown, Ed., Academic Press, New York, 1983.

Hayakawa, M., Onda, T., Tanaka, T., and Tsujii, K., *Langmuir* **13:** 3595–3597 (1997).

Hiemenz, P. C. and Rajagopalan, R., *Principles of Colloid and Surface Chemistry,* 3rd ed., chs. 10, 11 and 13, Marcel Dekker, New York, 1997.

Higashi, N., Kajiyama, T., Kunitake, T., Prass, W., Ringsdorf, H., and Takahara, A., *Macromolecules* **20:** 29–33 (1987).

Hoar, T. P. and Schulman, J. H., *Nature,* **152:** 102–103 (1943).

Hoffmann, H., *Structure and Flow in Surfactant Solutions—ACS Symposium Series 578,* pp. 2–31, C. A. Herb and R. K. Prud'homme, Eds., American Chemical Society, Washington, DC, 1994.

Hoffmann, H. and Ebert, G., *Angew. Chem. Int. Ed. Engl.* **27:** 902–912 (1988).

Hoshino, E., *Hyomen* (Surfaces) **29:** 709–723 (in Japanese) (1991).

Hoshino, E. and Ito, S., *Enzymes in Detergency—Surfactant Science Series 69,* ch. 9, J. H. van Ee, O. Misset, and E. J. Baas, Eds., Marcel Dekker, New York, 1997.

Huo, Q., Margolese, D. I., Ciesla, U., Feng, P., Gier, T. E., Sieger, P., Leon, R., Petroff, P. M., Schuth, F., and Stucky, G. D., *Nature* **368:** 317–321 (1994).

Ikeda, S, *Bull. Chem. Soc., Jpn.* **50:** 1403–1408 (1977).

Imokawa, G., Akasaki, S., Hattori, H., and Yoshizuka, N., *J. Invest. Dermatol.* **87:** 758–761 (1986).

Imokawa, G., Kuno, H., and Kawai, M,. *J. Invest. Dermatol.* **96:** 845–851 (1991).

Inagaki, S., Fukushima, Y., and Kuroda, K., *J. Chem. Soc., Chem. Commun.* 680–682 (1993).

Inagaki, S., Koiwai, A., Suzuki, N., Fukushima, Y., and Kuroda, K., *Bull. Chem. Soc., Jpn.* **69:** 1449–1457 (1996).

Ise, N., *Angew. Chem. Int. Ed. Engl.* **25:** 323–334 (1986).

Israelachvili, J. N., *J. Colloid Interface Sci.* **44:** 259–272 (1973).

Israelachvili, J. N., *Intermolecular and Surface Forces,* 2nd ed., Academic Press, New York, 1991.

Israelachvili, J. N. and Adams, G. E., *J. Chem. Soc., Faraday Trans. I* **74:** 975–1001 (1978).

Israelachvili, J. N. and Pashley, R. M., *Nature* **300:** 341–342 (1982).

Israelachvili, J. N., Chen, Y.-L., and Yoshizawa, H., *J. Adh. Sci. Tech.* **8:** 1231–1249 (1994).

Israelachvili, J. N., McGuiggan, P. M., and Homola, A. M., *Science* **240:** 189–191 (1988).

Israelachvili, J. N., Mitchell, D. J., and Ninham, B. W., *J. Chem. Soc., Faraday Trans. II* **72:** 1525–1568 (1976).

Israelachvili, J. N., Mitchell, D. J., and Ninham, B. W., *Biochim. Biophys. Acta* **470:** 185–201 (1977).

Ivanov, I. B. and Kralchevsky, P. A., *Thin Liquid Films—Surfactant Science Series 29,* ch. 2, I. B.Ivanov, Ed., Marcel Dekker, New York, 1988.

Jones, M. N., *Surface Activity of Proteins,* ch. 8, S. Magdassi, Ed., Marcel Dekker, New York, 1996.

Kamiya, T., Kawamata, A., Takaishi, N., and Hara, K., *Proceedings of the Japanese-US Congress of Pharmaceutical Sciences,* Honolulu, Hawaii, Poster Abstracts S273 (N 04-W-13); US Patent 4859696 and 4948588 (1987).

Kamo, N., Miyake, M., Kurihara, K., and Kobatake, Y., *Biochim. Biophys. Acta* **367:** 1–10 (1974a).

Kamo, N., Miyake, M., Kurihara, K., and Kobatake, Y,. *Biochim. Biophys. Acta* **367:** 11–23 (1974b).

Kanai, H. and Amari, T., *Rheol. Acta* **32:** 539–549 (1993).

Kauzmann, W, *Adv. Protein Chem.* **14:** 1–63 (1959).

Kawamata, A., Kamiya, T., Takaishi, N., and Hara, K., *Proceedings of the Japanese-US Congress of Pharmaceutical Sciences,* Honolulu, Hawaii, Poster Abstracts S275 (N 04-W-19) (1987).

Klein, J., *J. Colloid Interface Sci.* **111:** 305–313 (1986).

Klein, R. A., Moore, M. J., and Smith, M. W., *Biochim. Biophys. Acta* **233:** 420–433 (1971).

Kleinberg, M. E. and Finkelstein, A., *J. Membr. Biol.* **80:** 257–269 (1984).

Kobuke, Y., Ueda, K., and Sokabe, M., *J. Am. Chem. Soc.* **114:** 7618–7622 (1992).

Kodama, M. and Seki, S., *Adv. Colloid Interface Sci.* **35:** 1–30 (1991).

Kolp, D. G., Laughlin, R. G., Krause, F. P., and Zimmerer, R. E., *J. Phys. Chem.* **67:** 51–55 (1963).

Kose, A. and Hachisu, S., *J. Colloid Interface Sci.* **46:** 460–469 (1974).

Kose, A., Ozaki, M., Takano, K., Kobayashi, Y., and Hachisu, S., *J. Colloid Interface Sci.* **44:** 330–338 (1973).

Krafft, F,. *Ber.* **29:** 1334–1344 (1896).

Krafft, F., *Ber.* **32:** 1596–1608 (1899).

Krafft, F. and Wiglow, H., *Ber.* **28:** 2566–2573 (1895).

Kralchevsky, P. A., Nikolov, A. D., Wasan, D. T., and Ivanov, I. B., *Langmuir* **6:** 1180–1189 (1990).

Kralchevsky, P. A., Danov, K. D., and Ivanov, I. B., *Foams—Surfactant Science Series 57,* ch. 1, R. K. Prud'homme and S. A. Khan, Eds., Marcel Dekker, New York, 1996.

Kremer, J. M. H., van der Esker, M. W. J., Pathmamanoharan, C., and Wiersema, P. H., *Biochemistry* **16:** 3932–3935 (1977).

Kresge, C. T., Leonovicz, M. E., Roth, W. J., Vartuli, J. C., and Beck, J. S., *Nature* **359:** 710–712 (1992).

Kulkarni, V. S., Matsumoto, M., Yoshimura, H., and Nagayama, K., *J. Colloid Interface Sci.* **144:** 586–590 (1991).

Kumacheva, E., Klein, J., Pincus, P., and Fetters, L. J., *Macromolecules* **26:** 6477–6482 (1993).

Kung, H. C. and Goddard, E. D., *J. Phys. Chem.* **67:** 1965–1969 (1963).

Kunieda, H. and Shinoda, K., *J. Phys. Chem.* **82:** 1710–1714 (1978).

Kunieda, H., Akamura, K., and Evans, D. F,. *J. Am. Chem. Soc.* **113:** 1051–1052 (1991).

Kunieda, H., Nakamura, K., Olsson, U., and Lindman, B., *J. Phys. Chem.* **97:** 9525–9531 (1993).

Kunieda, H., Kanei, N., Uemoto, A., and Tobita, I., *Langmuir* **10:** 4006–4011 (1994).

Kunitake, T. and Okahata, Y., *J. Am. Chem. Soc.* **99:** 3860–3861 (1977).

Kunitake, T., Tsuge, A., and Nakashima, N., *Chem. Lett.* 1783–1786 (1984).

Kurihara, K. and Kunitake, T,. *J. Am. Chem. Soc.* **114:** 10927–10933 (1992).

Kurihara, K., Kunitake, T., Higashi, N., and Niwa, M., *Langmuir* **8:** 2087–2089 (1992).

Lange, H., *Solvent Properties of Surfactant Solutions—Surfactant Science Series 2,* ch. 4, K. Shinoda, Ed., Marcel Dekker, New York, 1967.

Lange, H. and Beck, K. H., *Kolloid Z. Z. Polymere* **251:** 424–431 (1973).

Lange, H. and Schwuger, M. J,. *Kolloid Z. Z. Polymere* **223:** 145–149 (1968).

Langmuir, I., *J. Am. Chem. Soc.* **38:** 2221–2295 (1916).

Langmuir, I., *J. Am. Chem. Soc.* **39:** 1848–1906 (1917).

Larsson, K. and Friberg, S. E., *Food Emulsions,* 2nd ed., Marcel Dekker, New York, 1990.

Lindblom, G. and Rilfors, L., *Biochim. Biophys. Acta* **988:** 221–256 (1989).

Lindblom, G. and Rilfors, L., *Adv. Colloid Interface Sci.* **41:** 101–125 (1992).

Loeb, S. and Sourirajan, S., *Sea Water Research,* Department of Engineering, University of California at Los Angeles, Report No. **58**-65 and **59**-3, 1958.

Longmuir, K. J., *Current Topics in Membrane and Transport: Membrane Structure and Function,* Vol. 29, p. 129, R. D. Klausner, C. Kempf, and J. van Renswoude, Eds., Academic Press, New York, 1987.

Lucassen, J., *J. Phys. Chem.* **70:** 1824–1830 (1966).

Luzzati, P. V., Mustacchi, H., Skoulios, A., and Husson, F., *Acta Crystallogr.* **13:** 660–667 (1960).

Mahler, H. R. and Cordes, E. H., *Biological Chemistry,* 2nd ed., pp. 714–724, Harper & Row, New York, 1971.

Malmsten, M., Claesson, P. M., Pezron, E., and Pezron, I., *Langmuir* **6:** 1572–1578 (1990).

Mandelbrot, B. B., *The Fractal Geometry of Nature,* Freeman, New York, 1982.

Mazur, A. and Harrow, B., *Textbook of Biochemistry,* 10th ed., pp. 90–94, Saunders, Philadelphia and London, 1971.

Miller, D. D., Bellare, J. R., Evans, D. F., Talmon, Y., and Ninham, B. W., *J. Phys. Chem.* **91:** 674–685 (1987).

Mino, J., Matijevic, E., and Meites, L., *J. Phys. Chem.* **80:** 366–369 (1976).

Mitchell, D. J. and Ninham, B. W,. *J. Chem. Soc., Faraday Trans. II* **77:** 601–629 (1981).

Moore, W. J., *Physical Chemistry,* 3rd ed., p. 738, Prentice-Hall, Englewood Cliffs, NJ, 1962.

Moroi, Y., *Micelles—Theoretical and Applied Aspects,* ch. 4, Plenum Press, New York, 1992.

Mukerjee, P. and Mysels, K. J., *Critical Micelle Concentrations of Aqueous Surfactant Systems,* NSRDS-NBS 36, U.S. Department of Commerce, Washington, DC, 1971.

Mukerjee, P. and Yang, A. Y. S., *J. Phys. Chem.* **80:** 1388–1390 (1976).

Muramatsu, M., Tajima, K., Iwahashi, M., and Nukina, K., *J. Colloid Interface Sci.* **43:** 499–510 (1973).

Murata, M., Hoshino, E., Yokosuka, M., and Suzuki, A., *J. Am. Oil Chem. Soc.* **68:** 553–558 (1991).

Murata, M., Hoshino, E., Yokosuka, M., and Suzuki, A., *J. Am. Oil Chem. Soc.* **70:** 53–58 (1993).

Murray, R. C. and Hartley, G. S., *Trans. Faraday Soc.* **31:** 183–189 (1935).

Naitoh, K., Ishii, Y., and Tsujii, K., *J. Phys. Chem.* **95:** 7915–7918 (1991).

Nakagawa, T. and Shinoda, K., *Colloidal Surfactants,* p. 160, K. Shinoda, T. Nakagawa, B. Tamamushi, and T. Isemura, Eds., Academic Press, New York, 1963.

Nakamura, M., Tsujii, K., Katsuragi, Y., Kurihara, K., and Sunamoto, J., *Biochem. Biophys. Res. Commun.* **201:** 415–422 (1994).

Napper, D. H., *Polymeric Stabilization of Colloidal Dispersions,* Academic Press, New York, 1983.

Nemethy, G. and Scheraga, H. A., *J. Chem. Phys.* **36:** 3382–3400; 3401–3417 (1962a).

Nemethy, G. and Scheraga, H. A., *J. Phys. Chem.* **66:** 1773–1789 (1962b).

Nikolov, A. D., Kralchevsky, P. A., Ivanov, I.B., and Wasan, D. T., *J. Colloid Interface Sci.* **133:** 13–22 (1989).

Ninham, B. W., Evans, D. F., and Wei, G. J., *J. Phys. Chem.* **87:** 5020–5025 (1983).

Nomura, T. and Kurihara, K., *Biochemistry* **26:** 6135–6140 (1987a).

Nomura, T. and Kurihara, K., *Biochemistry* **26:** 6141–6145 (1987b).

Nozaki, Y., Lasic, D. D., Tanford. C., and Reynolds, J. A., *Science* **217:** 366–367 (1982).

Oetter, G. and Hoffmann, H., *Colloids and Surfaces* **38:** 225–250 (1989).

Okahata, Y. and En-na, G., *J. Chem. Soc., Chem. Commun.* 1365–1367 (1987).

Okahata, Y. and Shimizu, O., *Langmuir* **3:** 1171–1172 (1987).

Okahata, Y., Ebato, H., and Taguchi, K., *J. Chem. Soc., Chem. Commun.* 1363–1365 (1987).

Okahata, Y., Lim, H.-J., Nakamura, G., and Hachiya, S., *J. Am. Chem. Soc.* **105:** 4855–4859 (1983).

Okumura, Y., Ishitobi, M., Sobel, M., Akiyoshi, K., and Sunamoto, J., *Biochim. Biophys. Acta* **1194:** 335–340 (1994).

Onda, T., Shibuichi, S., Satoh, N., and Tsujii, K., *Langmuir* **12:** 2125–2127 (1996).

Onsager, L., *Phys. Rev.* **62:** 558 (1942).

Onsager, L., *Ann. NY Acad. Sci.* **51:** 627–659 (1949).

Ozeki, S., Tsunoda, M., and Ikeda, S., *J. Colloid Interface Sci.* **64:** 28–35 (1978).

Padday, J. F., *Surface and Colloid Science,* Vol.1, p. 92, E. Matijevic, Ed., Wiley Interscience, New York, 1969.

Pashley, R. M. and Israelachvili, J. N., *J. Colloid Interface Sci.* **101:** 511–523 (1984).

Pashley, R. M., McGuiggan, P. M., Ninham, B. W., and Evans, D. F., *Science* **229:** 1088–1089 (1985).

Pethica, B. A., *J. Colloid Interface Sci.* **62:** 567–569 (1977).

Prince, L. M., *Microemulsions,* L. M. Prince, Ed., Academic Press, New York, 1977.

Princen, H. M., *Surface and Colloid Science,* Vol. 2 p. 61, E. Matijevic, Ed., Wiley Interscience, New York, 1969.

Privalov, P. L. and Gill, S. J., *Pure & Appl. Chem.* **61:** 1097–1104 (1989).

Reynolds, J. A. and Tanford, C., *Proc. Natl. Acad. Sci. USA* **66:** 1002–1007 (1970).

Robb, I. D., *Microemulsions,* I. D. Robb, Ed., Plenum Press, New York, 1982.

Robbins, C. R., *Chemical and Physical Behavior of Human Hair,* 3rd ed., ch. 1, Springer-Verlag, New York, 1994.

Roberts, G. G., *Adv. Phys.* **34:** 475–512 (1985).

Roberts, G. G., *Langmuir-Blodgett Films,* ch. 7, G. G. Roberts, Ed., Plenum Press, New York, 1990.

Rosano, H. L. and Clausse, M., *Microemulsion Systems—Surfactant Science Series 24,* H. L. Rosano and M. Clausse, Eds., Marcel Dekker, New York, 1987.

Rosen, M. J., Friedman, D.. and Gross, M., *J. Phys. Chem.* **68:** 3219–3225 (1964).

Rosensweig, R. E., *Adv. Electronics Electron Phys.* **48:** 103–199 (1979).

Sato, T. and Ruch, R., *Stabilization of Colloidal Dispersions by Polymer Adsorption—Surfactant Science Series 9,* Marcel Dekker, New York, 1980.

Satoh, N. and Tsujii, K., *J. Phys. Chem.* **91:** 6629–6632 (1987).

Scheludko, A., Toshev, B. V., and Bojadjiev, D. T., *J. Chem. Soc., Faraday Trans., I* **72:** 2815–2828 (1976).

Schieren, H., Rudolph, S., Finkerstein, M., Coleman, P., and Weissmann, G., *Biochim. Biophys. Acta* **542:** 137–153 (1978).

Schulman, J. H., Matalon, R., and Cohen, M., *Discussions Faraday Soc.* **11:** 117–121 (1951).

Schulman, J. H., Stoeckenius, W., and Prince, L. M., *J. Phys. Chem.* **63:** 1677–1680 (1959).

Shah, D. O., *J. Colloid Interface Sci.* **32:** 577–583 (1970).

Sherman, P,. *Emulsion Science,* P. Sherman, Ed., Academic Press, London, 1968.

Shibata, R., Noguchi, T., Sato, T., Akiyoshi, K., Sunamoto, J., Shiku, H., and Nakayama, E., *Int. J. Cancer* **48:** 434–442 (1991).

Shibuichi, S., Onda, T., Satoh, N., and Tsujii, K., *J. Phys. Chem.* **100:** 19512–19517 (1996).

Shinoda, K., *Colloidal Surfactants,* ch. 1, K. Shinoda, T. Nakagawa, B. Tamamushi, and T. Isemura, Eds., Academic Press, New York, 1963.

Shinoda, K., *J. Phys. Chem.* **81:** 1300–1302 (1977).

Shinoda, K. and Becher, P., *Principles of Solution and Solubility,* ch.7, Marcel Dekker, New York, 1978a.

Shinoda, K. and Becher, P,. *Principles of Solution and Solubility,* ch. 10, Marcel Dekker, New York, 1978b.

Shinoda, K. and Friberg, S. E., *Adv. Colloid Interface Sci.* **4:** 281–300 (1975).

Shinoda, K. and Hutchinson, E., *J. Phys. Chem.* **66:** 577–582 (1962).

Shinoda, K. and Kunieda, H., *J. Colloid Interface Sci.* **42:** 381–387 (1973).

Shinoda, K. and Lindman, B., *Langmuir* **3:** 135–149 (1987).

Shinoda, K. and Nomura, T,. *J. Phys. Chem.* **84:** 365–369 (1980).

Shinoda, K. and Soda, T., *J. Phys. Chem.* **67:** 2072–2074 (1963).

Shinoda, K., Hato, M., and Hayashi, T., *J. Phys. Chem.* **76:** 909–914 (1972).

Shinoda, K., Kobayashi, M., and Yamaguchi, N., *J. Phys. Chem.* **91:** 5292–5294 (1987).

Singer, S. J. and Nicolson, G. L., *Science* **175:** 720–731 (1972).

Sjöblom, J,. *Emulsions and Emulsion Stability—Surfactant Science Series 61,* J. Sjöblom, Ed., Marcel Dekker, New York, 1996.

Small, D. M., *J. Am. Oil Chem. Soc.* **45:** 108–119 (1968).

Small, D. M., *J. Colloid Interface Sci.* **58:** 581–602 (1977).

Smith, W. V. and Ewart, R. H., *J. Chem. Phys.* **16:** 592–599 (1948).

Sogami, I. and Ise, N., *J. Chem. Phys.* **81:** 6320–6332 (1984).

Sourirajan, S., *Reverse Osmosis,* ch. 1, Academic Press, New York, 1970.

Steinhardt, J. and Reynolds, J. A., *Multiple Equilibria in Proteins,* Academic Press, New York, 1969.

Stoeckenius, W., Schulman, J. H., and Prince L. M., *Kolloid Z. Z. Polymere* **169:** 170–180 (1960).

Sunamoto, J., *Farumashia* **21:** 1229–1235 (in Japanese) (1985).

Sunamoto, J., Mori, Y., and Sato, T., *Proc. Jpn. Acad.* **68(B):** 69–74 (1992a).

Sunamoto, J., Goto, M., Iwamoto, K., Kondo, H., and Sato, T., *Biochim. Biophys. Acta* **1024:** 209–219 (1990a).

Sunamoto, J., Sato, T., Hirota, M., Fukushima, K., Hiratani, K., and Hara, K., *Biochim. Biophys. Acta* **898:** 323–330 (1987).

Sunamoto, J., Noguchi, T., Sato, T., Akiyoshi, K., Shibata, R., Nakayama, E., and Shiku, H., *J. Contr. Release* **20:** 143–153 (1992b).

Sunamoto, J., Akiyoshi, K., Goto, M., Noguchi, T., Sato, T., Nakayama, E., Shibata, R., and Shiku, H., *Enzyme Eng.* **613:** 116–127 (1990b).

Suzuki, K., Okumura, Y., Sato, T., and Sunamoto, J., *Proc. Jpn. Acad.* **71(B):** 93–97 (1995).

Suzuki, T., Imokawa, G., and Kawamata, A,. *Nippon Kagaku Kaishi (J. Jpn. Chem. Soc.)* 1107–1117 (in Japanese) (1993).

Suzuki, T., Takei, H., and Yamazaki, S., *J. Colloid Interface Sci.* **129:** 491–500 (1989).

Suzuki, T., Tsutsumi, H., and Ishida, A., *J. Disp. Sci. Tech.* **5:** 119–141 (1984).

Suzuki, T., Nakamura, M., Sumida, H., and Shigeta, A., *J. Soc. Cosmet. Chem.* **43:** 21–36 (1992).

Suzuki, Y. and Tsutsumi, H., *Yukagaku (J. Jpn. Oil Chem. Soc.)* **36:** 588–593 (in Japanese) (1987).

Tajima, K., *Bull. Chem. Soc., Jpn.* **44:** 1767–1771 (1971).

Tajima, K., Nakamura, A., and Tsutsui, T,. *Bull. Chem. Soc., Jpn.* **52:** 2060–2063 (1979).

Takagi, T., Tsujii, K., and Shirahama, K., *J. Biochem.* **77:** 939–947 (1975).

Talmon, Y., Evans, D. F., and Ninham, B. W., *Science* **221:** 1047–1048 (1983).

Tanford, C., *J. Phys. Chem.* **76:** 3020–3024 (1972).

Tanford, C., *The Hydrophobic Effect: Formation of Micelles and Biological Membranes,* 2nd ed., ch. 1, Wiley Interscience, New York, 1980a.

Tanford, C., *The Hydrophobic Effect: Formation of Micelles and Biological Membranes,* 2nd ed., ch. 8, Wiley Interscience, New York, 1980b.

Tanford, C., *The Hydrophobic Effect: Formation of Micelles and Biological Membranes,* 2nd ed., ch. 14, Wiley Interscience, New York, 1980c.

Tokiwa, F. and Tsujii, K., *J. Colloid Interface Sci.* **41:** 343–349 (1972).

Tran, C. D., Klahn, P. L., Romero, A., and Fendler, J. H., *J. Am. Chem. Soc.* **100:** 1622–1624 (1978).

Traube, I., *Liebigs Ann.* **265:** 27–55 (1891).

Tsuchiya, M., Tsujii, K., Maki, K., and Tanaka, T., *J. Phys. Chem.* **98:** 6187–6194 (1994).

Tsujii, K. and Mino, J., *J. Phys. Chem.* **82:** 1610–1614 (1978).

Tsujii, K. and Satoh, N., *Organized Solutions: Surfactants in Science and Technology—Surfactant Science Series 44,* ch. 23, S. E. Friberg and B. Lindman, Eds., Marcel Dekker, New York, 1992.

Tsujii, K. and Tokiwa, F., *J. Am. Oil Chem. Soc.* **54:** 585–586 (1977).

Tsujii, K., Okahashi, K., and Takeuchi, T,. *J. Phys. Chem.* **86:** 1437–1441 (1982).

Tsujii, K., Saito, N., and Takeuchi, T., *J. Phys. Chem.* **84:** 2287–2291 (1980).

Tsujii, K., Hayakawa, M., Onda, T., and Tanaka, T., *Macromolecules* **30:** 7397–7402 (1997a).

Tsujii, K., Yamamoto, T., Onda, T., and Shibuichi, S., *Angew. Chem. Int. Ed. English* **36:** 1011–1012 (1997b).

Ueno, M., Tanford, C., and Reynolds, J. A., *Biochemistry* **23:** 3070–3076 (1984).

Udin, H., *Metal Interfaces,* p. 114, American Society of Metals, 1952.

Verwey, E. J. W. and Overbeek, J. Th. G., *Theory of the Stability of Lyophobic Colloids,* Elsevier, Amsterdam, 1948.

Yamamoto, T., Satoh, N., Onda, T., and Tsujii, K., *Langmuir* **12:** 3143–3150 (1996).
Yoshimura, H., Matsumoto, M. Endo, S., and Nagayama, K., *Ultramicroscopy* **32:** 265–274 (1990).
Yoshizawa, H. and Israelachvili, J. N., *Thin Solid Films* **246:** 71–76 (1994).
Yoshizawa, H., McGuiggan, P. M., and Israelachvili, J. N., *Science* **259:** 1305–1308 (1993).
Zumbuehl, O. and Weder, H. G., *Biochim. Biophys. Acta* **640:** 252–262 (1981).

Index

M

N

O

P